The present study is the result of a pilot project of the Development Centre on the impact of the Chinese economy in the Pacific Region and beyond. Research was initiated in late 1983 and involved the co-operation of several Chinese institutions and government agencies.

Also available

AGRICULTURE IN CHINA. Prospects for Production and Trade (July 1985)
(51 85 04 1) ISBN 92-64-12741-0   84 pages   £6.50   US$13.00   F65.00   DM29.00

JAPAN'S GENERAL TRADING COMPANIES. Merchants of Economic Development
by Kiyoshi Kojima and Terutomo Ozawa (January 1984)
(41 84 07 1) ISBN 92-64-12644-9   120 pages   £6.50   US$13.00   F65.00   DM29.00

NEW FORMS OF INTERNATIONAL INVESTMENT IN DEVELOPING COUN-TRIES by Charles Oman
(41 84 02 1) ISBN 92-64-12590-6   140 pages   £6.50   US$13.00   F65.00   DM30.00

DEVELOPMENT CENTRE STUDIES

CHINA'S SPECIAL ECONOMIC ZONES

BY
MICHAEL OBORNE

DEVELOPMENT CENTRE
OF THE ORGANISATION FOR ECONOMIC CO-OPERATION AND DEVELOPMENT

Pursuant to article 1 of the Convention signed in Paris on 14th December, 1960, and which came into force on 30th September, 1961, the Organisation for Economic Co-operation and Development (OECD) shall promote policies designed:

- to achieve the highest sustainable economic growth and employment and a rising standard of living in Member countries, while maintaining financial stability, and thus to contribute to the development of the world economy;
- to contribute to sound economic expansion in Member as well as non-member countries in the process of economic development; and
- to contribute to the expansion of world trade on a multilateral, non-discriminatory basis in accordance with international obligations.

The Signatories of the Convention on the OECD are Austria, Belgium, Canada, Denmark, France, the Federal Republic of Germany, Greece, Iceland, Ireland, Italy, Luxembourg, the Netherlands, Norway, Portugal, Spain, Sweden, Switzerland, Turkey, the United Kingdom and the United States. The following countries acceded subsequently to this Convention (the dates are those on which the instruments of accession were deposited): Japan (28th April, 1964), Finland (28th January, 1969), Australia (7th June, 1971) and New Zealand (29th May, 1973).

The Socialist Federal Republic of Yugoslavia takes part in certain work of the OECD (agreement of 28th October, 1961).

The Development Centre of the Organisation for Economic Co-operation and Development was established by decision of the OECD Council on 23rd October, 1962.

The purpose of the Centre is to bring together the knowledge and experience available in Member countries of both economic development and the formulation and execution of general policies of economic aid; to adapt such knowledge and experience to the actual needs of countries or regions in the process of development and to put the results at the disposal of the countries by appropriate means.

The Centre has a special and autonomous position within the OECD which enables it to enjoy scientific independence in the execution of its task. Nevertheless, the Centre can draw upon the experience and knowledge available in the OECD in the development field.

Publié en français sous le titre :

LES ZONES ÉCONOMIQUES SPÉCIALES
DE LA CHINE

TABLE OF CONTENTS

ACKNOWLEDGEMENTS

The present study would have been difficult to complete without the comments and suggestions of many China experts in both OECD Member countries and in the People's Republic of China.

The author would particularly like to thank the following individuals for their help in producing this study: Nicolas Fourt, for his indispensable technical assistance in collecting and analysing much of the primary data for Chapters 2 and 3; Frederic Bianquis for his work on investment problems in China; Ouyang Chen-shin and Ho Pu Yin for their work on Chinese materials.

PREFACE

The Special Economic Zones (SEZs) were part of the initial wave of enthusiasm for economic and political reforms that swept through China in the late 1970s. They were conceived as an instrument – small, specialised and limited in scope – to bring needed skills, technology and foreign capital into China in order to achieve the ambitious programme of the Four Modernizations. The zones, which are described in detail in the present study, were part of an on-going tradition of renewal and reform in modern China. Their success or failure was not to be the measure of the People's Republic of China's (PRC's) ability to confront the problems of the end of the century; rather, the SEZs were to be laboratories in which different doses of market economy planning could be experimented upon and adapted and imported into the socialist economy of China. The term laboratory is germane: laboratories are sealed off from the general public; special conditions prevail in them to ensure isolation of experiments. So too was it with the SEZs. Born of a desire to diversify the management and production system of China, the zones have received extraordinary powers and privileges. They have direct rule and a line of communications with the central government that allows them to obtain audience without delay for their special needs and problems; they have special political and administrative structures. The SEZs are enclaves of foreign presence and custom. Not unlike the efforts of late Qing China to modernize, the zones are called "open doors" onto the international economy.

The leadership in Beijing met the challenge of rapid modernization with a panoply of reforms and re-organisations of the economy. Those measures are beginning to be well known, now that teams from international organisations, government agencies and private businesses have travelled to China to study the conditions under which international co-operation can take place. The SEZs have their special place in the saga of the move to modernize the PRC in the 1970s. It is difficult to recall what China was like a brief ten years ago: shut to the outside world, deprived of information, unable to supply the most elementary statistics of past or current performance. Foreign presence was limited to contacts that the Chinese leadership deemed necessary; foreign businesses were not allowed to deal directly with Chinese counterparts.

By the end of 1984, a number of the experiments described in the present study had already been extended to inland enterprises in China. Export enterprises within the PRC were granted duty free conditions for raw materials being used for production (April 1983); the 100 per cent foreign-owned enterprises, an innovation in the Chinese socialist economy was at first limited to the SEZs, but by October 1983, this type of investment was permitted in "coastal cities where conditions permit". Management reforms, including the reform of the bonus system, the introduction of the piece-work bonus and more flexible hiring and firing possibilities, all of which were innovations of the SEZs, are now spreading to inland enterprises. Streamlined administrative supervision by the provincial and central government has also become a goal of the government. Foreign banks, which have limited branch office capacity within the SEZs, may soon be allowed to open commercial operations in other coastal cities. The greatest step in the direction of extending the experiment of the SEZs to a larger area inside China came with the announcement in April 1984 that 14 coastal cities would have special investment incentives for foreign joint ventures. The continued development of the SEZs is part of China's experimental programme to adopt new strategies for growth. The demonstration effect of these zones upon the rest of China is an important element in those growth strategies. How successful the SEZs will ultimately be depends a great deal on the manner in which the country as a whole can use these restricted zones as doors through which new ideas, technologies and goods can flow into China.

In recent months, there has been more and more criticism aimed at the SEZ experience. Some of this criticism was issued in the form of academic studies or journalist accounts of the Shenzhen SEZ appearing in the local Hong Kong press. Other criticism came from within China itself. Deng Xiaoping noted at the end of June 1985 that the SEZs were "experiments" and as such they might fail. If so, China would have to learn from that experience. He felt that they were not meant to be models for the country at large. Despite the negative side effects of the SEZs – the large consumer goods trade with the PRC which distorted production priorities for the zones, the rise of corruption and criminality and the general balance of payments difficulties of Shenzhen – Chinese authorities continue to support the experiment, although they seem more sanguine about short-term gains given the large initial investment in infrastructure. It appears that as long as the SEZs continue to play the vital role of testing ground for new management, production and commercial reforms their existence will not be seriously threatened. But that role is one with high exposure to wrong turns taken. The SEZs have already had a significant impact upon the Chinese economy, and as such have a lasting place in the strategy of Chinese development. Their story is a condensed version of the open door policy.

Just Faaland
President
OECD Development Centre
September 1985

Chapter 1

THE POLICY CONTEXT OF ECONOMIC REFORM
IN CHINA: 1977-1984

The post-Mao period in China is the subject of a good deal of research, within both government policy institutes and specialized university research departments. A short sketch of the policy background for economic reform in China is a necessary prelude to a discussion of the Special Economic Zones in Guangdong and Fujian Provinces, as these latter are conceived as tools for implementing the new economic policies. No attempt will be made to duplicate the detailed literature on policy evolution, but rather to set that research in a useful relationship to the present study.

China watchers throughout the world have been producing detailed studies of the rise of Deng Xiaoping to power and the subsequent espousal of a series of policies dubbed "pragmatic", "open door policies" by observers. Much of the post-Mao struggle for power has now been elucidated, and through official Chinese publications, scholars have a relatively coherent view of the political battles and victories won by the moderate faction of Deng Xiaoping after the fall of the "Gang of Four" in 1976[1].

For present purposes, our concern begins with the policies of early 1978 which the Chinese government announced as a means of accomplishing the Four Modernizations[2]. This later concept was in fact an old slogan used by Zhou Enlai as early as 1964 to rally China to a programme of rapid industrialisation. The four modernizations were to take place in industry, agriculture, science and technology, and defence[3]. After his rehabilitation in 1977, Deng promoted the modernization programme, once again showing his bent for practical economic planning by the use of capitalist techniques (especially entrepreneurial instincts) and incentives; thus he followed in good time the Leninist injunction to learn from capitalist practices and temper socialist dogma with a dose of pragmatic planning. The course of that programme of modernization is the framework within which the present study sets itself. The economic overhaul that the country underwent in the years 1978-1980 was a significant statement of the ability of China to re-orient its development programme by both reforming the internal economic system and opening the country to foreign influences and capital. The motivation for this series of changes is clear: the present leadership saw no viable alternative course towards growth within the present political system. Agricultural performance had to be improved to meet new needs and a growing population[4]; industrial productivity had to be increased without significant capital investment in new facilities, and more efficient use of the labour force had to be ensured[5]; the growing "under-employment" problem in both the rural and urban sectors did not bode well for economic and political stability[6]; Chinese use of capital investment has been notoriously poor over the past twenty years, and few new production techniques have been invented or introduced in China since the 1950s. With a labour surplus, the Chinese have tended to supplement high-cost capital equipment with workers, thus slowing down production and causing delays in standardized, serial production. Most of all, management and planning were not co-ordinated efficiently, and hierarchical management techniques, linked to party structures, made the production apparatus itself subject to the vagaries of political ideology. In the interim period of 1976, the process of decentralization went far beyond what the government had thought feasible, and a certain competition became the norm among provincial level industries who set targets on unit years instead of within the Five Year Plan; Beijing authorities had to regain control at the central government level of planning procedures.

There was a growing awareness among technical personnel in the government that China had fallen far behind developed countries in technology. Estimates by some Western observers put the lag at five to thirty years behind developed countries[7]. The imperative to catch up in the technology race was part of the decision to open up China to foreign investment and trade. Linked to the problem of technology gap was the severe shortage of trained scientists and technicians. This was due in no small part to the Cultural Revolution (1966-1975).

THE NEW HORIZONS

In early 1978, the Beijing government authorities announced a series of measures meant to put China clearly on the road to modernization. The Ten Year Plan was published, heralding the most ambitious effort for rapid modernization ever undertaken by a developing country. The Plan had the look of another call to mass mobilization on the scale of the Great Leap Forward. It proved otherwise; the egalitarian and mass mobilization policies associated with Mao Zedong (and upheld by Hua Guofeng, then Secretary-General of the Chinese Communist Party) were soon downplayed for more pragmatic goals. First, these reformers shored up their new policies for more rational economic planning (albeit unrealistic from the production output point of view) with the publication of an important policy speech of Mao from 1956 that fitted their needs. Then, the team around Deng began consolidating political opinion in favour of the new economic policies[8]. This orientation was approved by the Party's Eleventh Congress in August 1977. A more significant statement came from Hua Guofeng (then Secretary-General of the Chinese Communist Party) in his address to the National People's Congress, the civil legislative body of the country, in February 1978. In his "Report on the Work of the Government", Hua not only outlined the targets for the next Five Year Plan (1976-1980), but called for a merging of the Fourth and the Fifth Five Year Plans into a Ten Year plan. He also gave targets for China's economic performance until the end of the century. It was certainly the work of the Deng team, and was probably initiated under the direction of Zhou Enlai in 1974, when Deng worked with him as his chief deputy for economic affairs. This general economic policy re-orientation was followed by statements for science and technology and education which followed in the same vein of renewal[9]. The above-mentioned speech of the influential Politburo member, Hu Qioamu ("Act in Accordance with Economic Laws, Step Up the Four Modernizations") was a powerful climax to this policy evolution. Hu called for a form of economic reform in China that would move the country in the direction of "market socialism" although he did not employ this term specifically; from the text of the speech it is evident that the lessons he wished China to learn from other world systems came basically from the capitalist market economies.

The goals of the Ten Year Plan and the first stage of modernization were ambitious, especially when compared to other Five Year Plans[10]. Total capital construction investment in the period 1978-85 was to be equal to the entire expenditure on capital construction in the previous 28 years. Total agricultural output was to increase by 4 to 5 per cent a year and the industrial sector output was to increase by more than 10 per cent a year. Consequently, GDP growth (adjusted Net Material Product figures) would be in the range of 8 to 9 per cent per year. The plan called for 120 major projects to receive special priority, most of them large-scale developments, requiring massive construction and investment efforts. Transport infrastructure was to be significantly upgraded, and major heavy industries would receive priority for investment (metallurgy – ferrous, non-ferrous – transport, fuels, electric power and raw materials). Steel was given a special priority, continuing a Chinese (and Stalinist) penchant for believing in the necessity of autonomous fundamental industries at the national level. Ten large steel and iron complexes were to be expanded or built. By 1985, China's steel production was targeted at 60 million tons, more than double the 1977 production figure (24 million tons). Hua indicated that when the 120 major projects had been completed "China (would) have 14 fairly strong and fairly rationally located industrial bases under the supervision of six major regions, integrated into a comprehensive national system"[11].

This energetic plan was viewed with scepticism by the China watch community abroad. Few felt that the country had the resources to achieve even some of the targets. There was little analysis on the part of Chinese government of the important interaction of economic sectors among themselves. The timetable was clearly too ambitious.

The motor for the plan was not to be the mass mobilization policies of the past, nor the austerity policies of the Maoists; the new plan clearly was based upon innovative incentive devices and a system of personal responsibility that would relay individual energies into a collective economy, not through coercion, but enlightened self-interest.

As part of the new policy, provinces were to become more independent in light industrial manufacturing, and they were urged to achieve self-sufficience within as short a time as possible for certain consumer goods. This marked the swing-back tendency towards decentralisation of economic decision making.

The population problem was also squarely addressed. The goal of less than 1 per cent annual population growth was set for the country by 1981 (as compared to 1.7 per cent for the period 1965-77).

This programme of modernization proved unwieldly; each sector of the economy was called upon to perform at maximum speed. High productivity quotas were set for every sector, and the presumption was that allocation would be carried out without difficulty. The high real growth rates (expressed in net material product terms) predicted for the national economy were 8 to 9 per cent for 1978-1985. China's performance to date suggests that this forecast was not far off the mark (see Table 1). Initial targets for the Ten Year Plan predicted that the value of agricultural output would increase at an average annual rate of 4 to 5 per cent, and that industrial production would rise at a rate of more than 10 per cent. The actual economic performance has proven to be quite close to the latter predictions[12].

The factors that were most likely to affect (positively or negatively) the Chinese economy in the transition period were far reaching in the scope of activities and impact. The increasing need to use industrial output for the agricultural sector was certainly a concern of leaders in planning departments in Beijing, and this meant mobilizing support of state enterprises. Co-ordination of agricultural reforms and industrial reforms would be difficult at the national level. This purely quantitative observation was linked to the announced policy of modernizing the agricultural sector of the economy, a process that would take large industrial input to accomplish in even a modest fashion. The resistance of vested interest groups in the political hierarchy was also seen as a potential obstacle to the rapid growth rates.

Table 1. RATES OF GROWTH TARGETED AND REALISED IN THE FIVE YEAR PLANS

%

	GOVI annual rate of increase			GOVA annual rate of increase	
	Targeted	Realised		Targeted	Realised
1st Plan (1953-1957)	19	15.1	L 9.6 H 22.7	5	5
2nd Plan (1958-1961)	n.a.	−6.0 (1958-1962)	L −3.3 H −6.3	n.a.	6 (1958-1962)
3rd Plan (1966-1970)	n.a.	9.4	L 6.9 H 11.7	n.a.	2.8
4th Plan (1971-1975)	n.a.	7.7	L 8.0 H 7.5	n.a.	4.2
5th Plan (1976-1980)	10	11.2	L 13.2 H 9.6	4	5.7
6th Plan (1981-1985)	4.5 L 5.0 H 3.0	9 (1981-1983)	L 7.0 H 10.9	4.5	7.9 (1983-1984)

Note 1: GOVI stands for Gross Output Value of Industry
        GOVA stands for Gross Output Value of Agriculture
        L   stands for Light industry
        H   stands for Heavy industry
Note 2: The various growth rate entries are official deflated figures.

The leadership took steps to ensure that the government would be reorganized, and later that the purely political posts (such as party posts) would be separated from administrative and civil service appointments.

Human resource deployment also became a major issue in the modernization programme. Training and allocation of manpower within state and collective enterprises was identified as a major factor in the success pattern for growth.

Overall rates of national investment and the priority allocation of investment into energy and transport sectors remained a constant concern of the leadership. Even after identifying these large areas for action, it was clear that the necessary political will had to be marshalled in order to carry out the targeted reforms in the programme.

The policy mix required to foster rapid, sustained growth became the principal subject of debate in China. The two perceived growth goals of the policy makers were clear enough: concern for quantitative input-output performance and a qualitative streamlining of the economic system itself to generate greater efficiency throughout. The evolution of policy over the next five years demonstrated that the Chinese leadership was prepared to tackle serious adjustment problems within the country, and it is in the light of these policies for modernization that the importance of the Special Economic Zones as a tool for growth takes on meaning.

The year 1978 proved to be a high point of policy optimism. In February, Hua delivered his "Report on the Work of the Government", in which he outlined the long-term goals and ambitions of the then-named Ten Year Plan. This policy statement was immediately undercut by an official communiqué[13] which adjusted downward many of the planned targets, especially in the agricultural sector. The focus of policy in the early part of 1979 began to be called "adjustment" policy by the leadership in Beijing (as opposed to reform)[14]. The tide had clearly turned by the time Hua Guofeng presented the second "Report on the Work of the Government" to the Fifth National People's Congress (NPC) on 18th June 1979[15]. Hua presented a detailed rethinking of the Ten Year plan, conceding that the initial euphoria over growth prospects was now over, and that a three-year period was necessary to "adjust, reconstruct, consolidate, and improve the national economy in order to bring it step-by-step into the orbit of sustained, proportionate, and high speed growth". The general orientation, in fact, meant a better growth equilibrium, adjusting the capital investment goals to a quickened growth in consumption. Light industry and agriculture were to be emphasized over heavy industry. There was much talk of "efficiency" and "reforming the whole economic system", while at the same time providing more details for the sectoral targets[16]. The principal aim was to decentralise the economic decision making; this soon proved to be difficult, due to the increased necessity to control the national budget more tightly. Hua Guofeng was quietly removed as the nation's Premier, principally because of his role in drawing up the overheated Ten Year Plan.

The new policies were presented by many senior commentators during the year. The new Premier, Zhao Zi Yang's on-the-job experience in Sichuan and Deng's explanatory speech on the future of economic reforms remain, however, the most important political guidelines for the new policies of adjustment and reform[17]. The first results of this effort were a drastic decrease in scheduled capital construction projects including the halting of the large Baoshan steel complex near Shanghai, a balance of payments adjustment by a strong cutback in foreign imports and a reduction of the number of heavy industrial plants, particularly those that were deemed to be inefficient. Energy and transport became the two key sectors of the economy for development; it was more widely recognized that these two sectors alone created the major bottlenecks to growth. For energy, the production slowdown was particularly important. Coal, oil and natural gas production have declined since 1979, and as crude oil accounted for more than 14 per cent of the total export earnings in 1980, it was imperative to find new ways of opening oil fields in the country to offset balance of payment deficits occasioned by imports of grain, and equipment goods. For this reason, foreign oil companies were invited to begin oil exploration off the Chinese coast, in an effort to locate useable fields. The long-term energy situation was perceived as sufficient[18], but the leadership was divided concerning the means of creating more efficient energy use, as the consumption pattern for all energy users is extremely wasteful in China.

China's leaders also saw the need for reforming the whole of the national production system, especially by overhauling the commune-brigade-team hierarchy. In order to promote better production methods and more rational consumption patterns, price reforms were seen as a necessary adjunct to any reform of the economic system. However, all these measures were to be co-ordinated, and introduced gradually, to ensure even progress.

It would appear, however, that the much sought after goal of reducing capital investment for heavy industry does not seem to have paid off to date (1984), as preliminary calculations for the Sixth Plan indicate[19], (see Table 1). Note that the projected growth rates for heavy and light industry are inverted in actual performance terms.

However, the power of the central government to impose reforms was countered both within the government itself and by the local provincial authorities who during the first days of the Deng regime were encouraged to devise local economic planning methods as well as to start making decisions. Once they had gained that capacity to conduct business at the local level, the central planning units were disrupted by independent planning and decision making, especially in the area of capital investment and production. The result was a new state policy to recentralise key economic planning within the reorganized State Planning Commission (1980).

15

The adjustment policies were rooted in a perceived need to proceed more slowly – and therefore more thoroughly – in economic reform. Contradictory goals sometimes emerged: increasing agricultural output by modernizing production meant a potential unemployment problem, generated by more efficient production techniques; the family level production unit (the commune system was not formally abandoned until 1983), meant that collective property management as well as collective decision making became more of a problem; more foreign trade could mean increased dependency on the outside world, argued some. The structural adjustment policies adopted under the rubric of "Readjustment" were principally aimed at rescuing the economy from the chaos of the previous ten years. But the leadership certainly saw patent dangers in maintaining a policy mix which favoured industrial growth at the expense of consumer goods. The upheavals of both the Great Leap Forward and the Cultural Revolution had reduced consumer spending to a minimum, and the sacrifices required of the civilian population were constant grievances against the Maoist leadership; more attention had to be given to bettering the everyday life of the population – and in particular the peasants – if the new economic policies were to succeed[20].

The most critical problem for the central government remained – and still remains – a balancing act of great political skill. While more individual and local level responsibility and co-operation are needed, it is impossible for a country the size of China to continue sustained growth if the communal economic system is severely disrupted. The leadership in Beijing realized that they had to use carefully the Maoist capital of discipline, self-sacrifice and absolute obedience to the political system if they wished to achieve development goals. That political capital is their chief resource in manipulating the policies of the country at the national level.

The Period 1980-1984

National economic performance in 1980-1981 slowed down; appropriate policies were adopted to adjust the economy to inflation and to monitor budget and current account deficits that resulted from opening too widely the door to imports, contracting too many foreign loans and loosening the grip on the local planning authorities. The two-pronged policy goal of implementing reform and adjustment was quietly downplayed, as the leadership enacted measures in all sectors of the economy to co-ordinate more balanced economic growth[21].

By the beginning of 1982, the "reform" ideas were back into policy statements. Sweeping changes were made in institutional structures, both at the national and at the local levels. The most significant restructuring came about in the central government. The number of ministries and commissions in the already overblown bureaucracy in Beijing was reduced from 98 to 52, and the State Planning Commission and the State Economic Commission were given more authority and control within the government. There seems even to have been a movement to reduce the number of civil servants within the local and provincial levels of the government. Certain decision-making powers, including investment and finance decisions, were handed over to provincial authorities. The local branches of foreign trade corporations, so central to the Chinese import-export business, were given wider powers to negotiate directly with foreign trading partners[22].

The most widespread changes were instituted in rural and urban production systems. In an effort to relieve the chronic shortages of basic consumer agricultural goods, the authorities began to emphasize the "production responsibility system" that had been introduced nationwide in 1979. In the rural context, this meant a virtual decollectivisation of agriculture. The production brigade-commune system was replaced by a contract system in which families were allowed to farm collective land on a contractual basis. The new production unit was obliged to deliver quotas of goods at official procurement prices to the local authorities, but the surplus of the production could be sold on the free markets which sprung up everywhere as a result. Recently, the contracts awarded to peasants for individual farming were extended from a period of one to three years to a fifteen year period[23], principally to allay the fears of families that there would be a swing back to the practice of collectivisation in the short term. It was difficult at first to convince many peasants and urban workers that the repressive times of the Cultural Revolution were over for good, and that there would not be a new wave of denunciations in the future. Many remembered with bitterness the period of 1966-1972 when private initiative was tantamount to treason.

Alongside the "household responsibility system" the government authorities permitted the use of private plots by individuals, the surplus of which could be sold on the local free markets[24]. Prices were determined by supply and demand. State procurement prices were increased to add to the incentives to provide more varied and better quality goods[25]. The responsibility system has been extended to other rural areas of the economy, including forestry, pisciculture and fruit orchards.

This system of responsibility was the basic tool with which the government hoped to underline the relationship between output and income, a crucial problem in educating the masses in more efficient economic organisation. Thus emphasis upon management reform and private incentive in the economy were the hallmarks of the pragmatic policies introduced by the Deng team.

The sudden increase in earning potential led to a sharp rise in the incomes of a segment of the rural population, and has created some problems[26]. One side effect has been the tendency of the local authorities to increase quotas when sideline production proves a tempting alternative to increasing overall production unit performance when reporting to central authorities. This has dissuaded private initiative at the source, as the output increment is bought at official procurement prices.

Urban reforms have gone more slowly, as they are linked to more complex forms of production and involve different types of production leeway for generating sideline activities. Reforms have concentrated on increasing incentives both at the individual and the enterprise level. Reforms for State enterprises were announced in October 1984, as a second stage of reform. It is difficult to estimate the actual impact of these reforms[27].

Efficiency has become the key word in the economic policy milieu. Management reform, production output increases, more coherent pricing systems, better marketing and sales outlets, a better understanding of domestic and international business cycle fluctuations and appropriate responses to it, are all part of the plan to create a streamlined, less bureaucratic structure to serve the planned command economy of socialist China. The circulation of information and the creation of information systems is an obvious need linked to better knowledge of the international economy as a whole.

The two priority sectors of the economy, in terms of capital investment expenditures, were to continue to be the transport systems and energy supply systems.

During the period 1980-1984, the government followed a prudent path of setting purposely low growth targets, while at the same time continuing to carry out basic policy reforms throughout the economy. In order to bring the Chinese economy into a balanced growth pattern – something that is necessary before large-scale industrialisation or modernization at a complex level can take place – the leadership implemented macroeconomic policies in the following areas:

- restructuring of procurement prices for agricultural goods;
- a continued use of rationing for allocation of resources;

- fiscal reforms at the enterprise level designed to both lower budget deficits and increase local decision making on investment;
- a restrictive monetary policy and limited credit expansion for domestic investment;
- an attempt to reduce capital construction costs and attempts to change the pattern of capital investment in fixed assets financed out of central and local authorities' budgets;
- the implementation of reforms in the area of foreign trade, including an internal settlement rate for exchange transactions, to stimulate exports and reduce imports.

The effects of these policies have been impressive. The share of investment in the net material product (NMP) fell from 36 per cent in 1978 to 28 per cent in 1981; there has been a noticeable shift from heavy industry to light industry, agricultural production has increased, and the terms of trade have stabilized. There has been a reduction in the budget deficit, inflation has been curbed, the commercial balance has moved from a Renminbi (Rmb) 3.1 billion deficit in 1979 to a Rmb 5.6 billion surplus in 1982. The rate of economic growth has been steady: GOVA (Gross Output Value of Agriculture) and GOVI (Gross Output Value of Industry) have grown at 7.9 per cent between 1978 and 1983 in real terms. It also appears as though the strategy to reduce net investment has succeeded, with a corresponding growth in consumption.

Increase in rural and urban incomes has been impressive, due in part to 1.3 per cent per annum population growth rate between 1978-1981.

The industrial sector performed well during the period 1980-84. The energy infrastructure has been the single most important bottleneck for the industrial sector. Energy production remained stable during the period, although there was a growth in industrial output, due to more efficient use of energy. The heavy industry sector is notoriously wasteful in energy use. As energy products are also among the most important exports of China (petroleum products account for more than 20 per cent of exports), China can ill-afford to divert needed export earnings to wasteful production plants that have inefficient energy systems; consequently, a conscious effort has been made to introduce incentives to save on energy use rather than increase supply. Domestic consumption of energy grew by less than 2 per cent (1979-1982) due to both the shift in industrial structure (more emphasis upon light industry which uses one fourth as much energy per unit of output as heavy industry) as well as more efficient use of existing energy sources.

The balance between heavy and light industry, so essential to a better growth of a large range of consumer goods production, is linked to both investment and energy allocation. If local authorities are left to re-invest earnings in fixed assets, they invariably do so in either high profit light industry (cigarettes or light consumer goods) or heavy industry. If it is the latter, the energy supply necessary to run the plant is often either not available, or only useable during certain hours of the day. The dislocation of planning for investment and energy supply has lead to a severe shortage of electric power in many industrial areas, resulting in undercapacity production in both light and heavy industry. It is obvious that a price reform in the energy sector is necessary to reflect better the scarcity of energy supplies. Such a reform would have serious consequences on the cost of living index however, as energy is a subsidised quantity for every household as well as for the industrial production system.

The fiscal policies of the central government also saw significant changes in the period 1981-84. In an effort to set up a better sharing of responsibility for production, the central economic authorities have changed the profit remittance system for state and collective enterprises. Previously, enterprises handed over profits at the end of the fiscal year, once production costs had been paid. Losses were assumed by central planning authorities, and profits were absorbed to be used elsewhere, according to a masterplan. On 1st June 1983, a new system of accounting was instituted. Enterprises henceforth would participate in a profit retention scheme. Under the new arrangement, enterprises moved to a profit-based accounting system under which they paid regular taxes to the central authorities for production profits. This system was designed to place more responsibility upon the local enterprise management, as well as to render them more aware of (and eventually responsible for) losses. The central authorities reiterated that they intended to phase out or consolidate enterprises that consistently showed losses. By the beginning of 1984, more than 90 per cent of all state-run enterprises were paying taxes. One of the deleterious effects of this, of course, was to reduce the central administration's revenues. The result has been a reduction in central state spending, notably in the area of capital construction. The subsidy bill for commodities and services, however, rose sharply as a proportion of the expenditure, due principally to the fact that the government wished to keep inflation down, and thus used subsidies to finance the increased procurement prices for agriculture rather than pass these costs on to consumers.

The balance of payments performance for the whole country has been in line with the announced policy of the government to open China to the outside world and to encourage foreign trade. The merchandise goods sector of foreign trade has grown the most. This is in part due to the direct trading procedures which have been instituted, allowing provincial level authorities, and occasionally enterprise level authorities, to initiate trade agreements with foreign partners; it is also due to the use of the internal settlement rate that favours exports and virtually taxes imports[28].

The key to restructuring the economy is, in the opinion of the present Chinese leadership, combining state planning measures with personal initiative. Most reforms are heading in the direction of giving more leeway to the local level enterprises in rewarding productive workers, and in promoting efficient management schemes. The carrot side of the problem is being studied carefully by China watchers; the stick side is less well known. It is an obvious problem for leaders to design a path to economic growth that is commensurate with Chinese socialist goals; but it must be pragmatic enough to perceive that the period of mass mobilization and ideological rhetoric has given way to a new moment in Chinese history. This new moment would not have been possible without the sacrifices and restructuring of the past; but it appears likely that the present team around Deng feels that China cannot afford to remain a prisoner of that past.

SPECIAL REFORMS AFFECTING INWARD AND OUTWARD LINKAGES BETWEEN THE PEOPLE'S REPUBLIC OF CHINA AND THE SPECIAL ECONOMIC ZONES

The Special Economic Zones (SEZs) were created as an integral part of the new open door policies.

Several areas of the policy reforms described above have a special impact upon the functioning of the SEZs of the PRC. The policy context for the SEZs therefore is an important element in attempting to evaluate the success of their performance and fixing the costs and benefits of the SEZs in the light of the general economic reforms now underway.

Fiscal Reforms

The fiscal reforms that have accompanied the opening of China to the outside world have as their goal a rationalization of the production system, rendering more coherent the investment programmes of the major industrial sectors, and a streamlining of the economic system at large. Three phases of industrial fiscal reforms correspond to the progressive liberalisation of the economy since 1978. By 1979, the central authorities had decided that more responsibility for losses had to be shifted to the enterprise level of management (but not to the provincial level). This was accomplished by setting up "economic mechanisms" that would automatically redistribute responsibility. The 6 600 enterprises which produced more than 60 per cent of the national industrial output and delivered more than 70 per cent of the government's profit earnings were placed on a scheme that allowed them to retain some proportion of profits to be distributed within the enterprise as bonuses and production incentive raises; they could also be used for investment in fixed capital assets. The enterprise became responsible for planning production and modernizing equipment and physical plant installations. The effect of this first stage of reform was to reduce drastically the central government's income from profit remittances (a decline of 40 per cent in 1980). The attendant problem of over-ambitious local level capital construction became apparent as well. Enterprises embarked upon production expansion programmes that ate up profits, engaged too many funds, and required help from the central government to complete programmes. The rise in workers' wages, due to the new bonus funds and production related increases, also caused sharp rises in the consumer price index. At the same time, enterprises were not responsible for downstream marketing, as the totality of their production was purchased by central government commercial agencies. This led to over production and chaotic planning and co-ordination among enterprises and central government authorities. The central government closed this first stage of profit remittance reform in mid-1980, with specific measures: price controls were introduced on consumer goods, government spending at all levels was curtailed, and investment in fixed assets was reduced. Enterprise bank assets were frozen and could not be employed for investment purposes without the express permission of the central government; this reform was completed with the reform of the Bank of China in 1983. The state government also put into effect a plan to use loans to enterprises for fixed investment, rather than the system of centralised subsidies. The loans were to be paid back with enterprise profits, which in turn depended upon enterprise performance.

These deflationary measures increased the current account deficit in the central government, and the performance for heavy industry continued to draw investment funds for heavy industrial projects.

By the spring of 1981, the central government had decided to continue with fiscal reforms at the enterprise level, in spite of the growing deficit problem. A new contractual system, called "profits contracts" was introduced. Under this new system, enterprises negotiated a "base figure" of production with central ministries. Above this production, the enterprise could retain some 50 to 100 per cent of the profits. As these figures were negotiated each year independently of past performance, the enterprises could purposely quote low base figures, and once these quotas were filled, produce for the enterprise itself, without accounting to the central government authorities. No market mechanisms were taken into account, and enterprises vied with one another for surplus quota markets, where better-than-procurement prices could be obtained by enterprises for goods in demand. In fact, a hybrid system of production came into being: enterprises negotiated production quotas delivered at official procurement prices; surplus production could then be sold at higher prices, directly to clients. However, enterprises that failed to meet their quotas were not subject to disciplinary action, nor was there an incentive for enterprises to respond to market pressures, and adjust production to demand. Unsold surplus of one year could be used as part of a base figure for the following year. Local authorities also interfered more often at the enterprise level as a result of the shift away from central government management. The key element in this stage of fiscal reform was the promulgation of the contract law among Chinese contractual parties. The "Economic Contracts Law of the People's Republic of China" was promulgated at the end of 1981, and became effective 1st July 1982. Contractual arrangements among Chinese parties were divided into ten "horizontal" categories: purchase, construction, product processing, transportation of goods, supply of electricity, storage, rental of property, loans, property insurance, and technological co-operation. The law sets the stage for the promulgation of a "vertical" law, binding partners to co-operation within the chain of production. The above law applies only to planning directives already approved by the central government. The government has announced that a law is in preparation involving foreign parties and Chinese enterprises *(Law on Economic Contracts Involving Foreign Parties)*[29].

In early 1983, a third major adjustment was made to the enterprise system. The "tax-instead-of-profits" scheme was extended to a large number of essential industries (it had been operative on an experimental basis since 1980 in 200 state enterprises). Under the new system, a tax would replace the former profit remittance system, and would involve charges on fixed and circulating capital, as well as a commercial tax; a flat 40 to 60 per cent tax was levied on declared profits. There would also be an adjustment tax levied on more profitable enterprises in favour of less profitable ones. This policy has received public support at the highest political levels[30].

This policy has greatly enhanced the power of the central government authorities, as it made transfer payments a matter of accounting. Solvency and profit margins now have become the concern of enterprises, as successful or failing business operations are much more visible.

The two areas left for future action at the enterprise level were the relationships with local political authorities, and the problem of the market oriented commercial strategies. Although no precise measures were adopted to phase out unprofitable enterprises, it was part of the government plan to render these more vulnerable, and to take steps to consolidate, or even shut them down.

The result of these fiscal reforms was a multi-level system of revenue sharing with the central government. In the most common of fiscal regimes still practised in China, the local and provincial government authorities share all local revenues with the central government. This includes profits from the enterprises that are not involved in the new tax-instead-of-profits system, as well as enterprises that still remit profits. It also includes income tax levied on foreigners living in China (this is not a large source of revenue for the moment), and taxes on banks. It does not include customs duties.

In 1982, some 80 per cent of local revenues were retained by provinces. The independent municipalities had a lower rate, an indication that they were subsidising the provinces (Beijing, 36.5 per cent retained; Tienjin, 31.2 per cent, Shanghai, 11.2 per cent).

A second revenue sharing involves dividing only the industrial and commercial taxes, but not other enterprise level profits. The province of Shensi currently falls into this category.

Guangdong province has a fixed rate of payments each year, rather than a percentage of profits. Fujian province also had this regime for some time but since 1982, fits into the following category: a fourth fiscal regime consisting of net transfers of revenue from the central government to the provinces (eleven such transfers in 1980).

Reforms in Capital Costs Allowances for Enterprises

The central government had to make the enterprise level aware of, and responsible for, eventual economic efficiency. Among other reforms introduced in 1979, a measure to require enterprises to bear fixed capital costs was introduced. Previously, enterprises applied to the central government for such costs; under the new system, the local governments were made responsible for collecting fixed capital cost payments from enterprises' declared profit shares. A typical breakdown of such costs is as follows: 5 per cent capital use fee, 55 per cent income tax, 15 per cent on fixed and circulating capital. This would leave an enterprise with approximately 25 per cent of profits.

The strategy of allowing local government agencies to plan and carry out investment with retained profits after taxes has not completely supressed the tendency of the local level officials to use profit for investment at the local level. The resulting competition for construction materials has caused a substantial increase in construction costs (9 per cent since 1978). The central government has only been able to complete 47 of the 80 large-scale projects that were originally scheduled for 1984.

Tax measures are the economic instrument with which the government wishes to enforce its policies for reform. In a large measure, the success of grass roots reform will depend upon the success of the new tax programme, as it is an automatism that will operate at all levels in favour of a more efficient and rational planning. By mid-1984, these fiscal policies had attracted a good deal of attention both inside and outside China[31].

Price Reforms and Subsidies

Chinese economic planners face related problems of pricing and subsidies. Recently, Prime Minister Zhao Ziyang targeted price restructuring as the single most important economic reform now facing China, echoing much of the literature on Chinese pricing practices that has been written in the past five years[32]. In the realm of prices, it is evident that substantial adjustments will have to be made in the planned prices of essential commodities, bringing the price into line with the relative scarcity of the good. At the same time, the market must be allowed to operate more freely, in order to ensure that goods are produced and sold in response to demand, with corresponding price incentives. Many of the basic raw materials now used for production in China are undervalued on the domestic market, distorting the fixed price for the processed material. It also allows the enterprise to register artificially high profit margins, due to the implicit subsidy in the purchase of component inputs of base elements. At the same time, a price reform of the basic type that is needed throughout the economy could easily bring on inflation, something the central government wishes to avoid at all costs, remembering the years after Second World War when inflation was one of the major problems in China.

Currently, some of the planned prices for essential commodities have been adjusted upward (including coal, cotton, iron, timber) and a few of the basic commodity prices have been allowed to float downward in response to supply. It is clear that a central planning authority has difficulty responding to price adjustments throughout the economy, and it is not until a clearer idea of the actual costs to the economy of specific production processes are calculated that the government can take vigorous enough steps to adjust[33].

The price mechanism is part of the set of policies being used to adjust production to demand. This also is carried out through the selective purchasing of goods by the central government. The commercial units of the government have been experimenting with refusing

22

to purchase surpluses that are deemed unsaleable; factories are left with the possibility of selling the remainder on the direct market at quasi-official prices or assuming the loss.

Related to price reforms is the issue of subsidies. The average urban worker receives substantial subsidies in food, housing, medical, and transport allowances. An estimated 13 per cent of the national income is devoted to subsidies; in terms of the individual worker's salary, it works out to approximately 89 per cent of his nominal wage. Subsidies for rural workers are much less important, and represent about one tenth of those accorded urban workers. There is, in fact, a subsidy gap which is increasingly favourable to the urban worker[34].

Economic Policy and Reform: The October 1984 Statement

While the reforms in the agricultural sector advanced with great speed during the period 1980-1984[35], urban and industrial reforms were much less well publicised. As a means of introducing many of the agricultural responsibility system reforms into the general economy, the government presented an ambitious blueprint for urban reforms in October 1984[36].

The thrust of the new urban reforms was restructuring the state and collective enterprise system[37]. The text of the decision to introduce urban reforms stresses that the central planning system, which was adopted after the 1949 revolution, was adequate for the early growth period of China, but was no longer adapted to the increasing complexity of the economy, especially in the industrial sector. The overemphasis upon the state-owned enterprise system had virtually edged out all private initiative in the economy, and as recent agricultural reforms had demonstrated, this was a valuable component in increasing both worker productivity and overall economic performance when properly tapped. The Beijing leaders clearly see the same type of flexibility necessary in the industrial sector, if dramatic gains are to be registered. In order to carry out enterprise reforms, however, it was evident that the production units themselves would have to be thoroughly overhauled. Management reform emerges in the text as the single most important enterprise level reform necessary for China. The series of reforms included a series of new accounting procedures, new fiscal regulations, more flexibility in recruitment and the fixing of workers wages, a supply and distribution system that is more attuned to market mechanisms and a general concern for efficiency within the enterprise.

Perhaps the most dramatic changes announced were in the area of price reforms. The price structure of the country had been distorted by a cumbersome subsidy system which favoured urban residents; it also made it virtually impossible to assess the profitability of enterprises; at the same time it caused high production of goods with large profit margins, and short supply of goods with small margins. The state budget was drastically affected by the subsidy system, and current account deficits caused by subsidies were common[38]. Serious efforts at reform were put off until the Seventh Five Year Plan (1986-1990). However, by June 1985, a few timid steps began the move towards more rational prices, including the raising of farm procurement prices. This latter move was made as incentives to agricultural production as well. As these costs were not passed on to consumers through higher retail prices, the subsidies accorded to farmers under state budgetary procedures rose. In areas where over-production persisted, prices were allowed to float downward, to reduce inventory and eventually discourage enterprise over-production. The majority of staple food products, however, as well as energy supply items and the majority of heavy industrial inputs remained under fixed price controls. Only the agricultural free market production was completely freed from any state control.

The 1984 Decision speeded up the price reform. In essence, it announced that price would in the future reflect more clearly production costs, and would take into account supply and demand. Although the mechanics of the price reform are not spelled out, it is clear from the document that prices will not be allowed to rise in a general fashion, nor will inflation be allowed to become a problem. With the gradual introduction of higher prices, "the various economic agents will be given time to adjust to the changes by altering their production or input mix, or by absorbing cost increases through productivity gains"[39]. Wages can be expected to rise, thus off-setting to some degree cost of living rises in consumer goods.

Planning in the economy will be subsequently modified. The 1984 Decision forsees guidance planning instead of regulatory planning for many sectors of the economy. Greater use will be made of indirect signals and tools for economic allocation of resources, such as correct price signals, fiscal policies, new wage scales, credit and profit rates and new methods for calculating enterprise profits. Mandatory planning will be largely limited to major products, and will be handled centrally:

> "The central plan will include output targets for energy, rolled steel, cement, basic raw materials for chemical industry, and synthetic fibre. In agriculture, targets will be used for cereals, cotton, edible oil, tobacco, jute, pigs, and fish products. All in all, mandatory planning will eventually pertain to only 60 industrial and 10 agricultural commodities, down from 120 and 29 at present."[40]

Management functions in enterprises will also be given more attention than in the past. Continuing in the line of weakening the party's hold on the economy, the Deng team has emphasized the need to separate government from enterprise functions, thus letting the local management take more responsibility for the profits and losses of the enterprise. It frees the government administration to spend more time monitoring the economy and adjusting, where necessary, the economic levers.

NOTES AND REFERENCES

1. See for instance the very complete review of the period 1976-1980 in Doak Barnett's *China's Economy in Global Perspective*, Washington D.C., 1981; T. Pairault's *Politique industrielle et industrialisation en Chine*, Paris, October 1983; and the well documented chronicles in the *China Acktuel*, Hamburg, (monthly) and *The China Quarterly*, London, (quarterly). Chinese sources include Deng Xiaoping's own statements recorded in official press releases, as well as numerous publications in Chinese newspapers (*Renmin Ribao* particularly).

2. For a comprehensive review of the economic policies of the Four Modernizations, see the Selected Papers of the Joint Economic Committee, US Congress : *China Under the Four Modernizations*. Part I, 1979; and Parts II and III, August 1982, Washington D.C.

3. Zhou had called for these at the Third NPC in December 1964 and again in January 1975 at the Fourth People's Congress.

4. It was reported in an article of the *Renmin Ribao*, 6th October 1978 that the average per capita grain distribution in 1977 only matched that of 1958. Hu Qiaomu "Act in Accordance with Economic Laws, Step up the Four Modernizations".

5. Hu Qioamu stated that growth in productivity was merely due to the greater number of workers added each year to the force, and not to any greater efficiency. There were also capital equipment additions, of course.

6. President Li Xiannian, then Vice-Premier, was reported to have stated in 1979 that 20 million people were unemployed in China. This would mean that about one fifth of the urban work force were unemployed. Reported in *Ming Bao*, Hong Kong, and the *Washington Post*, 15th June 1979 in the article "Chinese Official sees Economic Crisis".

7. See *Barnett, op. cit.*, p. 32; also *China Under the Four Modernizations, op. cit.*

8. Mao's 1956 speech entitled "On the Ten Major Relationships" was delivered to a Politburo meeting. The text was not published at the time. A new official version was presented in *Peking Review*, No. 1 (1977), pp. 10-25. It differs from versions smuggled out of China in 1956. Many of the positions taken by Mao in his later life do not resemble those he presented in the 1950s. The Deng team exploited this to show at the same time that Mao was inconsistent at the end of his life and that they were following "true Maoist thought" in introducing reforms.

9. Fang Yi "Outline of the National Plan for the Development of Science and Technology: Relevant Policies and Measures" *Peking Review*, No. 14, (7th April 1978), pp. 6-7; and Deng's speech on education as reported in *Peking Review*, No. 18, (5th May 1978), pp. 6-12. Both are important policy statements on the subjects treated.

10. See especially D. Barnett, *op. cit.;* A. Eckstein, Quantitative Measures of China's Economic Output, Ann Arbor, 1980. For an overview of some of the reforms, see Yu Guangyuan (ed.), China's Socialist Modernization, Beijing, 1984.

11. Hua Guofeng, "Unite and Strive to Build a Modern Powerful Socialist Country : A Report on the Work of the Government", 26th February 1978, as reported in *Foreign Broadcasting International Service,* (FBIS) *Daily Report-PRC,* 7th March 1978, D16.

12. Annex 1 for industrial output, agricultural outpout for the period 1978-1983.

13. Text in *FBIS Daily Report-PRC,* Supplement 015, 2nd July 1979, pp. 1-32.

14. See especially "Some Questions Concerning the Acceleration of Agricultural Development", text in English in *FBIS, Daily Report-PRC,* Supplement 123, 25th October 1979 and "Draft Proposals Concerning Some Problems in Speeding Up the Development of Industry", text published in *Beijing Review*, No. 28, 14th July 1979. This text is the "Thirty Point" strategy, an updating of Deng's "Twenty Point" strategy for industry. Industrial strategy is reviewed carefully in T. Pairault's *Politique industrielle et industrialisation en Chine,* Paris, 1983.

15. Text in *FBIS Daily Report-PRC* Supplement 015, 2nd July 1979, pp. 1-32.

16. See Hu Qiaomu's important policy statement "Act in Accordance with Economic Laws, Step up the Four Modernizations", in *Renmin Ribao* 6th October 1978 (in Chinese); English text in *FBIS, Daily Report-PRC,* 11th October 1978, E1-E21.

17. Zhao's experiments in Sichuan province where he was First Secretary of the Sichuan Provincial Party Committee (1975-1981) were outlined in his speech entitled "Study New Conditions and Implement the Principle of Readjustment in an all-around Way", published in Chinese in *Honggi*, No. 1, 1980. Deng's speech was reprinted in English in *FBIS, Daily Report-PRC,* 11th March 1980, U1-27.

18. China has known coal reserves of 600 billion tons, 1.8 billion tons of oil and 130 billion cubic meters of natural gas.

19. For an overview of the real performance of the Sixth Five Year Plan, see *Beijing Review,* 16th September 1985, p. 14 ff.

20. For a comparison of the new economic policies with other socialist efforts, see Y. Chevrier "Chine: Que veut Deng Xiaoping", *Politiques Étrangères,* No. 1, 1985; for an excellent overview of the readjustment period, see Jean-Philippe Béja, "Chine: 35 ans, ça suffit?", *Esprit,* April 1985.

21. For a complete overview of the economic performance of the PRC during the first years of the reform and readjustment periods (1978-1980), see the *World Bank Country Study: China, Socialist Economic Development,* Washington, D.C., World Bank, 1981. (Three volumes).

22. See discussion of these bodies in Chapter II of the present study.

23. As announced in "Party Document on Rural Work", *Beijing Review,* 20th February 1984, p. 6.

24. It is estimated that approximately 10 per cent of all rural households are now engaged in this contract system, called "specialised households". More than 90 per cent of production teams have some form of responsibility system operative. Cf. *Beijing Review.*

25. Grain, cotton and oil-bearing crops still must be sold to state market organisations, but procurement prices for the latter have risen 23 per cent in 1979, 7 per cent in 1980 and 6 per cent in 1981.

26. Crime has become a serious problem in some areas. Crop theft and livestock rustling are not uncommon in more affluent areas. Some production units have created special crop watch functions to protect produce destined for the free market.

27. The reforms were announced in an important policy statement made to the NPC on 20th October 1984. The full English text was presented in *Beijing Information,* 29th October 1984.

28. See Chapter 2 on internal settlement rate.

29. See *State Council Decrees,* Art. 11, 13th December 1981.

30. Prime Minister Zhao Ziyang reiterated the desire of Beijing to implement the tax reform at the enterprise level before the 1984 Session of the NPC (May 1984). He stressed the need to "do away with egalitarianism in industry and encourage competition" through a restruction of the fiscal system, as reported in the *Financial Times,* 16th May 1984. At the same time, it was announced that although directors of major enterprises will still be appointed on political criteria, the directors will now be free to appoint managers and carry out major enterprise-wide changes without reference to the central government. As reported in *Le Monde,* 24th May 1984, p. 3.

31. For a complete overview of the tax reforms from a Chinese point of view, see the article of Wang Chuanlun "Some Notes on Tax Reform" in the *China Quarterly,* No. 97, March 1984. This article reviews a number of articles in Chinese academic journals on the problem of enterprise tax reforms and economic planning.

32. See the three volume World Bank Study on China cited above for more details on this subject. See also the November-December 1983 issue of the *China Business Review,* and most particularly the article of Barry Naughton, "The Profit System", p. 14 ff. In a recent address to the NPC, in May 1984, Zhao Ziyang expressed the opinion that the pricing system was "irrational" and that it needed urgent reform. This is the opinion of the majority of economists who follow Chinese development.

33. It is reported that the Chinese economic planning authorities have begun computing an input-output table for 30 sectors of the economy and that this should be ready by 1986. It is to be followed by a thorough price reform based upon the results. See Naughton, *op. cit.* p. 18. A short series of articles on prices was published in *Beijing Review,* August 1983, No. 35-36, which review policies pursued by the government through 1983. Premier Zhao has announced a thorough price reform for China to begin in 1985.

34. Lardy, N. *Agricultural Prices in China,* September 1983, World Bank Working Paper, 1983 and article in *China Business Review,* November-December 1983.

35. For a review of the agricultural reforms, see in particular: Lin Zili, "On the Production Responsibility System Which Links Income to Output by Production Quota Contracts", *Zhongguo shehui kexue,* No. 6, 1982; A. Watson, "The New Structures in the Organisation of Chinese Agriculture: A Variable Model", in *Pacific Affairs,* No. 57, Winter 1984-1985, pp. 621 ff; and a rather general presentation in *Beijing Review,* Vol. 27, Nos. 44, 45, 46, 50, 1984.

36. "Decision of the Central Committee of the Communist Party of China on Reform of the Economic System", text in *Beijing Review,* Vol. 27, No. 44, 1984.

37. For a review of the economic reforms announced in 1984, see L. de Wulf, "Economic Reform in China", *Finance and Development,* March 1985 and the March-April issue of *China Business Review.* The *China Economic News* also presents detailed documentation on the implementation of economic policy, and the translation of all important policy statements.

38. See N. Lardy, "Runaway Subsidies", *China Business Review,* November-December 1983.

39. De Wulf, *op. cit.,* p. 10.

40. See "Decision", 1984, *op. cit., Beijing Review,* section V "Establish a Rational Price System and Pay Full Attention to Economic Levers"; and de Wulf, *op. cit.,* p. 10.

Chapter 2

CHINA'S FOREIGN TRADE AND INVESTMENT POLICIES AND THE SEZS

The development policies of the Chinese government in the period 1980-1984 have evolved rapidly, and at times, in contradictory fashions; they have been chronicled and documented elsewhere in detailed research[1]. For the purposes of the present study, the most important areas of policy interaction between the SEZs and the domestic economy relate to foreign trade and foreign direct investment, as the SEZs were established to promote both. These fields have also been the subject of recent scholarly work, although the lack of coherent data has hampered work on foreign direct investment[2]. Recently, the Chinese statistical officials have been providing more reliable data, however there is still no complete breakdown of foreign investment as opposed to barter trade agreements. In the latter, equity shares are calculated in currency for contractual purposes, but shares and profits are actually paid in capital equipment or production goods.

The following review of the foreign trade system and the foreign investment structures in the PRC provides the background for discussion of the SEZs as part of China's open door policy. In the quest for modernisation, the interrelated roles of foreign trade and foreign capital investment were highlighted. For example, by 1979, the Bank of China was promoting domestic trade related loans; to finance this, the Bank borrowed capital on the international market. It is reported that more than $19 billion was loaned by the Bank for domestic use during the period 1979-1983[3].

The structural reforms of the economy continued – on paper as well as in the field – throughout 1983-1984. The perceived need for greater foreign direct investment meant overhauling the foreign trade structures, as well as the foreign investment structures. These changes are detailed below in the context of a discussion of the foreign trade and investment reforms.

In the first instance, a review of China's traditional trade practices and attitudes towards foreign contact are reviewed, in order to place the SEZ experiment in the context of cyclic openings and closings of the Chinese economy.

CHINA'S FOREIGN TRADE IN PERSPECTIVE

Traditional Chinese attitudes towards outsiders governed foreign trade relations for more than fifteen centuries. The PRC has inherited many vestiges of that long imperial past. The SEZs of today are located in the same areas as those cities reserved for foreign contact and trade in the 19th century, as well as those traditional areas for trade and contact with Arab merchants under the Tang, Yuan and Ming dynasties. If "development" means something in a Chinese context, it is this conscious and unconscious primacy of continuity with a long past[4]. In explaining the mechanism of the SEZs of today, the roots of this experiment must be examined; without according the historical dimension too great a place in the argument, it is necessary to place the SEZs in the context of China's cycles of continuity and rupture in her long economic development.

Intimately linked to the problem of opening China to the outside world is the vision that the Chinese, and their successive rulers – both Chinese and foreign – have entertained concerning outsiders. The Chinese empire was an agrarian phenonemon, concentrated in the rich agricultural areas south of the Mongolian Steppe. The Great Wall, a concrete limit to Chinese expansion, demarcated the difference between the essentially pastoral economies of the plains and the tight-knit farming communities in the south. The Chinese, since the Han unification of the 3rd century BC, began to consider their culture and civilisation as unique, and in some ways superior, to others in their known world[5]. Their administration, technology, urban organisation, monetary and communications systems were more developed than any other contiguous nation. The Inner Asian tribes with which the Chinese had contact were rough and ready horsemen, unsettled in a given territory and a mobile menace to the Han civilisation. Of the two major cycles of Chinese history, that of dynastic change and that of foreign invasion, the second was the more prominent fear, as it often heralded the first. A set vocabulary and stereotype image of the barbarians developed in the highly hierarchic Confucian society. The non-Han barbarians were often associated with cardinal points of the compass, a concept that had profound effects upon the image of the Westerners who first arrived in China in the 16th century (they were associated with the barbarians of the East, the "I"). The Chinese developed a concept that has been contrasted with the Western notion of nationalism, "culturism"[6]. At the apex of the Confusian cosmology was the Son of Heaven, who through his virtuous action (Te) held the divine right of government over the Han (the so-called Mandate of Heaven). This concept of Chinese civilisation as the centre of the world, the emanation of all culture, was the demarcating factor between the known world and the barbarians. The latter were invited by the very ideology itself to "come and be transformed" by contact with the wise and perfect government of the Celestial Empire. The relationship between the emperor and the barbarians came to symbolise the actual historical relationship between China as the centre of culture and the rude tribes roundabout[7].

Inherent in this relationship with the foreign barbarians was the reciprocal notion of the generosity and compassion of the Son of Heaven, who would bestow the blessings of civilised contact, and at times political legitimacy, upon the outsiders who came to the Court. It was fitting that this balance of power should be embodied in some form of concrete exchange. Thus grew up the practice of what the Chinese found a natural relationship, the tribute of barbarians. The benefits for the tributary state were multiple, and such hommage to the Son of Heaven carried with it the right to exchange not only goods, but to become, in some sense, part of the Celestial Empire, the modern equivalent of being put on the map in the wake of independence[8]. The tribute system that touched most of the Far East in the period AD 700-1600 was both a medium for diplomatic discussion and negotiation and a means of sampling and exchanging goods from areas that were far from the Middle Kingdom. It was truly an international communications network; the Chinese court performed the function at times of arbiter and promoter of peace and stability in the whole of East Asia, excluding Japan. It has been recognized that at different times in Chinese history, the tribute system served different purposes, sometimes shoring up the failing dynasty, sometimes a means of expansion, and at all times a means of intelligence about the outside world[9].

The advantages of the tribute system were apparent for the Chinese. The tributaries themselves were sometimes no more than disguised merchants, under the name of ambassadors or diplomats. They brought with them the goods of their respective nations. Early in history the Chinese began to regulate this foreign trade associated with the practice of tribute. Emporia were set up at the frontier posts where the entourage of the ambassador could clear the goods duty free, and where goods could then be sold to retail merchants. Alternatively, the goods were carried to the capital and then sold at special markets set up outside the Residence for the Tributary Envoys. All of these transactions were carefully supervised by the Board of Revenue, the Ministry of Economy of the times. Independent merchants had to deposit goods at the frontier emporia. For the Central Asian tributaries, the emporia were located along the Mongolian borders, for the Koreans, at the Manchurian border. For those maritime traders and tributaries, the emporia were located in the city of Canton (Guangzhou), in the south of Guangdong Province. This city became the centre of the foreign sea transport trade, and it was into this provincial capital that the first substantial trade missions of Western Europeans came during the mid-16th century. It was entirely natural, from the Chinese point of view, that the Portuguese, and later the Dutch and English, should be relegated to this area, as it was here that much of the traditional maritime trade with Eastern barbarians had long been confined. It was also natural that the Chinese should confound the newcomers with their own notion of tributary-barbarian-trader, for so it had been in the secular memory of the Court[10].

The China coast had, of course, extensive contact with foreigners. Arab trading communities had existed since Tang times (AD 618-906) in the port of Zayton near modern Shanghai, and on the island of Amoy (Xiamen). The cities of Suzhou and Fuzhou were used as foreign contact points as well. The principal of sequestering foreigners in a special area and requiring that they should not take up permanent residence in China was an off-again-on-again policy of different dynasties. It seems to have been aimed at reducing the effects of foreign customs and mores upon the Chinese population, and in this regard, the Ming (1368-1644) restrictions on Chinese travel and the virtual closing of the empire to foreign travel are reminiscent of some of the policies of Mao in the 20th century. There is a strong tradition, therefore of shunning foreign contact in the heartland of China.

These foreign quarantine quarters were the forerunners of free trade zones, for duty was levied only on special categories of imports. More crucial, the foreign trade was in the hands of foreigners themselves.

This confusion of trade and tribute in the minds of the Chinese administration brought the country into head-on collision with changing times. The first serious blow to the isolation of China came from the contact with Jesuit missionaries who brought with them new scientific techniques and toys. The reform of the calendar, the introduction of longitudinal calculations, new astronomical techniques, timepieces and worst of all, new notions of cosmology disturbed and upset the time-honoured ideas of Chinese superiority. The famous incident of Matteo Ricci presenting the new European made world map to the Wan Li emperor reveals both the power of received ideas about foreigners and the role of China in the then known world (c. 1600). Trigault's account of the incident is surprisingly modern in tone, despite having been written in 1618:

> Of all the great nations, the Chinese had the least contact with outside nations, and consequently they are grossly ignorant of the world in general. True, they had charts ... that were supposed to represent the whole world, but their universe was limited to their own fifteen provinces, and in the sea painted around it, they had placed a few islands to which they gave the names of different kingdoms they had heard of ... it is evident, as they call their country Tienhia, meaning everything under the heavens; when they heard that

30

China was only a part of the great East, they considered such an idea so unlike their own, to be something utterly impossible ... The geographer (i.e. Ricci) was obliged to change his design (of the map he offered to the emperor) ... leaving a margin on either side of the map, making the kingdom of China to appear right in the centre. This was more in keeping with their ideas ...[11].

To this day, China is still called "the Middle Country" in the Chinese language (Zhong Guo).

Foreigners engaged in trade were confined to special limited contact points around the empire. But the Chinese merchants themselves also conducted wide trade among the tributary states. As the diplomatic tributary function declined, this external Chinese trade increased[12]. It was not until the late Qing that a distinction was made between trading countries and tributary countries, although the Imperial ministry that dealt with both was one and the same. The anachronistic attitude toward foreign trade within China was a sign that the imperial administration was unable, or unwilling, to face the realities of a changed world, one in which China was no longer at the centre, but a partner in the community of nations. Trade carried on by Chinese abroad was permitted under the Qing, but trade with foreigners inside China itself was forbidden, linked as this practice was with the age-old notion of tributary status. This situation stifled free exchange and irritated the newcomers, off-spring of commercial nations in Europe and the new world that depended upon trade to motor their own economic development. The clash of mentalities was inevitable. China was too rich a market, too important a political entity to remain aloof from the economic interdependence that was establishing itself in the world of the early 1800s.

The Canton System and the First Foreign Trade Zones

When Westerners arrived in China in the mid-16th century, they unknowingly found themselves in one of China's great down cycles of dynastic change. The Ming dynasty was about to give its last competent emperors; by 1644, the dynasty was obliterated by its own intransigence and decadence, to be replaced by a vigorous foreign, semi-Sinified dynasty, the Manchus (their dynastic reign title was Qing, 1644-1911). The Manchus, as foreigners themselves, had to rely upon the conquered Chinese administration to keep the country running; they created a dyarchy of civil and military orders in which each nationality had specific powers (the Chinese generally ran the civil administration, up to, but not including the Ministerial level, and the Manchus manned the army). This foreign dynastic presence – so odious to the Chinese modernists of the late 19th century – was an important element in the forced opening of China in the 1830s.

By the beginning of the 19th century, the principal elements of the foreign trading system were already in place: the civil administration provided special officials to oversee foreign trade in a few selected special locations along the coast (mainly at Canton and in Fujian province) the quarantine of foreigners to ports of call, and even specialised sections of the port where they were allowed to make short stays (this recalls the Japanese solution under the Tokugawa Shogunate of limiting foreigners to the island of Deshima in Nagasaki Bay from 1641-1868); a special tax rate was levied on all goods before they were brought into the country, generally 30 per cent of the estimated value. The workings of this complex system are the subject of much scholarly literature[13]. The cycles of foreign trade with China in modern times are interesting barometers to the internal social situation, and might be studied quantitatively when data become more readily available[14]. By the end of the 18th century, an elaborate system of special rules and regulations existed for foreign trade; it was especially well developed in the city of Canton. There, the chief superintendent of Maritime Customs, the notorious "Hoppo" of Western grievances, supervised taxation, and acted as a special go-between for the foreigner and the Chinese world. Special Chinese guilds were formed (the "hongs") and a collective merchant guild was created, the Cohong, to deal with Western trade at the wholesale level. Internal rivalries grew up among Chinese merchants; those who controlled foreign trade in the special coastal cities secured the issuance of imperial edicts that reiterated Qing hostility to foreign trade by Chinese abroad, and linked the home-based trade to the tributary system; thus the Hong merchants saw it in their own interest to maintain the fiction that contact with Europeans and the lucrative trade that ensued be inscribed under the ancient tributary system that brought the foreigners to China, and not a free trade system that would encourage the Chinese to become roving merchants.

The British East India Company, like the overseas trading companies of other European nations, had a monopoly on China trade granted by royal charter. These forerunners of the multinational companies carried out many of the same tasks in the past that transnational corporations perform today: they secured far flung markets for clients, ensured the smooth flow of goods by controlling as many of the stages of production, manufacture, transporting and retailing as possible, creating innovative banking and transfer payment systems as they went into new markets; they brought new ideas and technologies into economies with less economic verve. How successful they were in the 18th century remains to be assessed; they were curiously seen as enemies of the spirit of individual free enterprise, and with the rising tide of the industrial revolution in Britain, parliamentarians were brought under increasing pressure by the new economic order to break the monopoly of the "Company".

The East India Company "country trade", a triangular affair among Britain, India and China was one of the remote causes of change in the Chinese empire. The elaborate arrangements that the Company made with venture capitalists to further the flow of raw materials and semi-manufactured goods from India to China are much too complicated to detail here, and sufficient accounts have been given elsewhere of this phenomenon[15]. The Canton trade changed drastically in the first quarter of the 19th century, as privateers pressured the British government to sanction and protect the opium trade, which by now had become the pillar of the country trade. Tea and silk left China in large quantities; but woollens, the major British export, were impractical goods to sell in the semi-tropical climate of Southern China. As the terms of trade worsened for British merchants in the early 1800s, other goods were traded, including opium, which was produced in abundance in Bengal. The cycle by which supply produced in turn demand for the drug is a well documented preliminary to the Opium War of 1840. The bare fact remains that the conflict that came to be known as the First Opium War (1840-1842) was largely an attempt by foreign traders, including Western Europeans and North Americans, to open up Qing China to foreign trade.

At least some have seen the Anglo-Chinese struggle as a conflict of world views, a truly cultural war:

"Specifically, the Opium War represented a clash between two conceptions of international order. The Western system of national states clashed with the traditional Chinese view of a universal ethico-political order under the Son of Heaven. The two sides had conflicting economic conceptions. China's self-sufficiency and disesteem for the merchant led her to regard foreign trade as unimportant, not as a national economic necessity. Westerners could never understand why the Chinese government restricted its merchant trade. The Opium War had its immediate origin in a dispute over legal institutions. British ideas of the impersonality and supremacy of the established codes, their view of evidence and of legal responsibility came into direct conflict with the Chinese view that the emperor's administration should operate on an ethical basis above the mere letter of the legal regulations"[16].

The consequence of the first, and later a second, conflict between China and the Western powers was the establishment of the famous "treaty ports". This traumatic contact with the West on Chinese soil brought the Qing government to the painful realisation that it was no longer in a position to dictate to the "foreign barbarians"; the creation of the treaty port system modified forever the tributary system which had governed the political and economic contact with foreigners for the Celestial Empire. Henceforth, trade would be conducted on an equal basis with sovereign partners who did not acknowledge the superiority of Chinese culture, and whose interest in profit superseded their interest in ancient rights. The modifications to the trade system imposed upon the Qing dynasty coloured – and still colours – China's view of her place in the international community. Much of the current debate (including the Chinese refusals to recognise "privileges" of the foreigners in the zones) surrounding the SEZs has its roots in this conflict of interests in the treaty port settlements of the mid-1800s. It is important to understand the Chinese view of the treaty ports, and their reaction to the "opening of China" to the international economy in order to situate present day arguments in a context that makes sense.

The Treaty of Nanking (29th August 1842), ended the first round of hostilities between the Chinese and Western powers. It set out the new conditions for foreign trade within designated port-cities: the Cohong foreign trade monopoly in Canton was abolished, "fair and regular tariffs" were promised according to a set *published* schedule; the island of Hong Kong was ceded to Great Britain as a trade base, and five other Chinese ports were to be opened to British trade and residence: Canton (Guangzhou), Amoy (Xiamen), Foochow (Fuzhou), Ningpo (Ningbo) and Shanghai. An indemnity was also paid to the British government (21 million Mexican dollars). This treaty was supplemented by three more treaties (1843-1844), with the United Kingdom (the British Supplementary Treaty), the United States and the Government of France. These treaties established what was to be known as the "most favoured nation clause" which in practice meant that any privilege extracted from the Chinese by one power was automatically applied to all other signatories. In the end, the whole formed a single treaty system.

One of the most controversial aspects of these treaties was the extra-territoriality conferred upon foreign nationals living in the treaty ports. This stipulation meant that Chinese law did not apply to foreigners living in the treaty ports; they were under the jurisdiction of their consulates. This applied to property as well as persons. In some cases, it even applied to Chinese assistants, thus ensuring that Chinese associate merchants (and missionaries) were free from Imperial control within the treaty port itself.

As painful as this co-existence with foreigners was, the contact with the West proved to be fruitful: from it was born the first movements of modernisation in China, including the remarkably familiar concept of "Zi-chiang" (self-reliance, self-strengthening)[17]. Technology transfer became an interest of the forward looking Manchu and Chinese officials, and some saw these limited port contacts with the foreigners as an ideal opportunity to learn from the West, observing and absorbing new methods.

The Qing dynasty, however, remained impervious to the situation. Senior officials deeply resented the intrusion of foreign merchants and missionaries, now protected by foreign governments. When the Taiping Rebellion – a momentous peasant uprising and ensuing civil war fuelled by curious Christian hybrid ideologies – threatened the dynasty itself, it became evident to the rulers of China that the country had to strengthen itself against foreign merchants, even if it meant learning from them in a first stage.

As a result of foreign help given to the Chinese in putting down the Taiping Rebellion, the second treaty system became operative in the 1850s. The Western powers obtained further trading privileges; nine more ports were opened to foreign trade, and foreign vessels were permitted to engage in the transport of passengers and trade goods within China itself; although not mentioned explicitly in the treaties, the opium trade was legalized; foreign imports were allowed to pass freely to the interior after the payment of a duty of 7.5 per cent (duty plus transit), thus allowing strong competition with local production. More important, the Shanghai Foreign Inspectorate became the regulating agency for foreign trade within the Chinese Empire, and the Chinese Imperial Maritime Customs was created and staffed at senior levels with Westerners. Foreign trade became an exclusively foreign affair. The growth of Shanghai as a treaty port and a major industrial centre in the early 20th century owes much to this status of international city where capital could be attracted without the fear of Qing or later, Republican, interference.

The quest for modernisation in China is part of this story as well. The terrible impact of the continual wars and rebellions in the 19th century brought some educated Chinese, and even some Manchus, to the conclusion that China would have to acquire modern science and technology quickly and engage more actively in the international trading system.

The peregrinations of modernism in China are only beginning to be understood from a socio-economic point of view[18]. The parallel modernisation and entry into the international economic system of Japan poses the problem of why, with similar attitudes towards trade and foreign contact, Japan was able to respond with the Mejii Restoration, while China was unable to create even a moderate form of modern social and economic order[19]. The socio-political situations were quite different, it is true, but many Chinese intellectuals looked to Japan in the late 19th century as an Asian nation that was firmly on the road to economic development, and hence a model to be emulated. Japan joined the Western powers as an industrialising nation, and pursued its own policies of economic development without much concern for the declining Chinese empire's inability to deal with new forms of technology and international trade.

The expertise of the treaty ports was a vital element in the modernisation of China before 1940, as a specialist in Chinese development has pointed out:

"Two fundamental conditions were absent for economic growth in China: the emergence of a modern nation state, and a technological and institutional transformation in agriculture. Both were delayed despite the efforts of the modernizers in the 19th century and the Nationalist government in the 20th century. Therefore, the central government played virtually no role in the economic modernisation of the mainland until the 1930s. The principal agents of economic change and innovation were:

a) foreign enterprises and their native followers in the treaty ports;
b) foreign governments and their chosen instruments in Manchuria;
c) a few outstanding Chinese statesmen of the late 19th century;
d) the Nationalist government in the 1930s"[20].

The treaty port system lasted in China, in one form or another, well into the 20th century. World War II provided a convenient watershed for the concept, and most Western powers renounced the system themselves. When Mao Zedong came to power, he vowed to erase the memory of that humiliating chapter in Chinese history. The evolution of Shanghai from 1950 to this very day is a dramatic proof that the Communist regime meant to build a new China, effacing wherever possible the treaty port system and its privileges.

The treaty port presence of foreigners, and the privileges exacted from the Chinese government, left deep resentment among nationalists and intellectuals. The forced opening of China to foreign trade and foreign residence became the hallmark of China's decline. The Communist regime vowed to efface that humiliation by closing China once again to the outside world, and conducting a new exercise of self-strengthening and self-sufficiency. This autarchy was the official policy of the Chinese government from 1950-1978; with the exception of the Soviet technical mission in China during the 1950s, the Middle Kingdom once again was off-limits to foreign traders, investors and residents.

The creation of the SEZs is, then, something of a surprise. For these zones, which allowed foreign trade, foreign investment and foreign residence for the first time since the revolution, were located in areas that had been treaty ports (Shantou, Xiamen), or near enclaves wrested from the Chinese by foreign powers (Macao, Hong Kong). The new zones, on the other hand, would be very different from the old port cities: in the SEZs, Chinese law prevails, Chinese administrations run the local government, and Chinese investment matches foreign investment. The new legal systems set up to ensure the smooth functioning of the zones for foreign investors are not the extraterritorial law of yesteryear, nor are the zones to become carved out enclaves where wealthy Chinese and foreign entrepreneurs congregate. In short, the Chinese perceived a need to open in a cautious manner to the outside world, to ensure the acquisition of new technology, new management skills, and a means of bettering their trade performance; there were also demonstration effects created by the SEZs: the "one country, two systems" doctrine of Deng Xiaoping was illustrated by the success of Shenzhen; and the Hong Kong question was settled peacefully, in part due to the Chinese flexibility in introducing reforms into the SEZs that would act as "buffer" effects between the future special administration zone of Hong Kong and China. The balance of gains for China far surpasses, however, the mere creation of export processing bases in the traditional sense of balance of payment effects. The SEZs, faint shadows of the treaty ports, heirs to the export processing zone experience, are in fact unique Chinese creations, forging links outward to the international trading and finance systems, and forging links inward to the remoter Chinese provinces. The SEZs are doors through which pass goods and services into the country; they are also windows through which China acquires a better knowledge of the industrial world beyond her shores, and at the same time permits a privileged picture of China itself to foreigners. This amazing component of the open door policy is the vanguard of economic reform in China, and poses queries that are far more complex than the usual questions concerning export processing zones in developing countries.

The two most important economic activities that affect the SEZs are China's foreign trade and foreign investment. Before 1978, neither sector was important in the economic life of the country, reflecting well the desire to be self-sufficient and self-reliant. This notion of auto-sufficiency in trade was clearly part of China's difficulty in entering the foreign trade system in the 18th and 19th centuries, as several scholars have pointed out[21]. The open door policy brought dramatic changes in both areas, and it is no exaggeration to say that it was through these two economic activities that the opening of China to the outside world took place in the late 1970s. The SEZs were created, in part, to experiment with reforms in both the foreign trading system and the foreign investment code for the whole of China.

ORGANISATION OF FOREIGN TRADE
AND FOREIGN DIRECT INVESTMENT CHANNELS

In evaluating the role of the SEZs, the evolution of the foreign trade structures in China is important to recall, as it is the chronicle of how political authorities have attempted over the past years to implement different policies of decentralisation and modernisation. It also illustrates some of the problems faced by government officials in assessing the hidden effects of change. In the following discussion, the foreign trade activities have rather arbitrarily been separated from the foreign direct investment policies, although in the PRC these activities are closely related. This arises from the fact that a dominant form of foreign investment in China has been compensation agreements, where goods are exchanged rather than capital investment transfers.

Before the reform of the foreign trading system in 1979, the organisation of Chinese foreign trade followed closely that of the Soviet Union; almost all transactions were conducted by the Ministry of Foreign Trade (MOFT), and by the Foreign Trade Corporations (FTCs). FTCs carried out day-to-day foreign trade responsibilities in their product area. They are independent corporate bodies authorised to handle all business and legal problems in their area of expertise. They are supervised by the Ministry of Foreign Trade.

The MOFT played a key role in assuring the monopoly of foreign trade by carrying out the following tasks:[22]

- establishing, with the help of other ministries, the plan for foreign trade, as well as a plan for the use of foreign currency received as payments for goods;
- establishing contacts, on one hand, with potential clients and foreign producers, and on the other with local Chinese production units;
- verifying that all foreign trade transactions were compatible with national priorities, as set out in the Plan;
- signing contracts with foreign organisations, paying out or taking in foreign currency payments;
- regulating accounts with Chinese production units and aiding in the administration of the customs and quality control of products.

The import-export functions were divided into product categories; there were eight national FTCs:

1. China National Cereals, Oils, and Foodstuffs, Import-Export Corporation;
2. China National Native Produce and Animal By-Products Import-Export Corporation;
3. China National Textiles Import-Export Corporation;
4. China National Light Industrial Products Import-Export Corporation;
5. China National Chemicals Import-Export Corporation;
6. China National Machinery Import-Export Corporation;
7. China National Metals and Minerals Import-Export Corporation;
8. China National Technical Import-Export Corporation.

The national FTCs had local offices in municipalities, and regional offices for the more remote areas. These latter were controlled by the Regional Foreign Trade Bureau, an organism which depended directly upon the MOFT; there was little regional or provincial influence over these offices.

The organisations which the MOFT represented (production units with import or export business) were not directly associated with the negotiation of contracts but they did have the right to send observers to negotiation sessions. At the end of the 1970s, the central government decided to allow certain "production" ministries to open their own FTCs in order to profit from their concrete experience in producing and selling to the state. In part, this move was designed to pare down the growing bureaucratic structure of the central government. The first ministry to have benefited from this "horizontal" decentralisation was the First Ministry of Mechanical Construction, which created the Chinese National Corporation of Machines and Equipment (CNCME); the administration of the FTC was given over to a mixed board of officials from both the CNCME and the MOFT; the latter had discretionary veto rights over projects. By the end of 1980, 17 ministries had obtained the right to create their own specialised FTC. Other central organisations such as the China National Petroleum Corporation or the China State Shipbuilding Corporation (both directly under the State Council) received the same sort of authorisation.

In 1979, a cyclic large-scale decentralisation began. The trading practices of the MOFT were severely criticised by the new group of leaders around Deng Xiaoping. The new team of political planners saw the MOFT as an inefficient and overloaded bureaucracy, incapable of handling the enormous needs of the country in promoting exports and securing needed imports at good prices. The decentralisation that followed was more particularly aimed at the export sector. It associated the state enterprises more closely with negotiations, allowing them direct contact with foreign clients and encouraged state and collective enterprises to reduce delays in exporting goods. The reforms also modified the administrative procedures that were required by both Chinese and foreign partners in negotiating imports and exports. Perhaps most important, the local level enterprises were allowed access to foreign exchange earnings from their exports as an incentive to conduct business in a more commercial fashion. It was not until late 1984, however, that the central government instituted special privileges for the local branches of the Bank of China to facilitate this local use of foreign exchange.

The principal ideological reason given for the reform was the rationalisation of the socialist system, and the move away from authorative central government action at every level of economic development[23]. As a result, a gradual shift from central FTC to local FTC activity took place. In 1980, the local FTCs were managing 2 per cent of total import trade, and 8 per cent of export trade; by 1981, they handled 8 per cent of import trade and 13 per cent of export trade

The reform was institutionalised by the creation of a new ministry, the Ministry of Foreign Economic Trade and Relations (MOFERT), which took over the functions of the Ministry of Foreign Trade, the Ministry of Economic Relations with Foreign Countries (essentially an aid agency), the Import-Export Commission (an organisation that was consulted in the case of very large FDI projects and turnkey operations), and the Foreign Investment Control Commission. The MOFERT was given the following functions:

- propose to the State Council the drafts of laws and regulations concerning foreign trade;
- supervise the FTCs;
- supervise the customs, and fixes customs tariffs;
- supervise the bureau of "quality control" for foreign goods;
- carry out business and commercial negotiations with third countries, and sign agreements and protocols in the purview of the central government;
- authorise access to foreign exchange reserve to domestic organisations;
- study and subsequently approve all foreign investments in China, including those from Hong Kong and Taiwan, as well as all Chinese investments that surpass a ceiling set by the central government; this does not hold true for the SEZs, nor the 14 coastal cities and selected inland cities;
- co-ordinate plans concerning foreign trade production ministries with central planning authorities, and regional planning authorities in view of establishing a national plan for foreign trade;
- negotiate directly the import and export of certain strategic materials;
- grant import/export licenses for selected raw materials (notably energy related products) and goods;
- fix the price of exported goods; in the case of goods for which the MOFERT is directly responsible, prices are fixed at MOFERT; for other products, the MOFERT

Figure 1.   ORGANISATION OF FOREIGN TRADE IN 1979

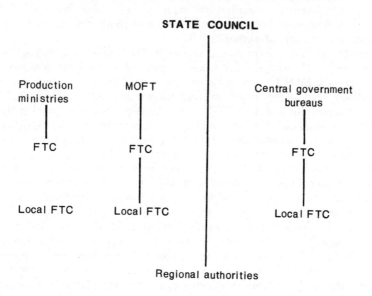

publishes a price schedule to avoid undue competition among Chinese organisations selling similar goods. The MOFERT therefore controls the profit margin of enterprises exporting the majority of their production.

Thus, the MOFERT, even if it does not have the decision monopoly for trade, controlled the all-important FTCs. This was a source of conflict between the regional authorities and the MOFERT, as the officials of the former MOFT never really accepted the reduction in their power. Figure 1 presents an organisational chart (1979) which emphasizes the role of the MOFT and its FTCs, as well as the "production" ministries and certain central government offices that are authorised to carry out trade directly with foreign countries via their own FTCs. Figure 2 presents the simplified structure of foreign trade as reorganised in 1983; it also represents the major channels of transaction for the country. There is now a Provincial Office of Import and Export which depends upon both provincial level administrations and the MOFERT. This office has the power to review all trade relations at the provincial level. Even if it does not direct the FTCs which depend directly on the MOFERT, nor the specialised FTCs, it has an overview of all trade transactions, and carries out trade in product areas that are not covered by the MOFERT.

In October 1984, new decentralisation policies were announced for the foreign trade system. The new reforms aim at two interrelated issues: rendering the production enterprises more responsible for export performance and further decentralising the foreign trade administration. Under the new system, which is scheduled to be set up gradually in 1985-1986, production enterprises will have a direct hand in export business. It appears that export enterprises will be created alongside production enterprises; these new service enterprises will be responsible for exporting goods. The new system is called an "agent" system, and corresponds to the new powers afforded the state enterprises which make of them accounting units as well as production units. As part of the reform, 64 foreign trade bodies replaced the eight original FTCs[24]. By mid-1985, more than 1 000 direct trade corporations were created.

Figure 2.   ORGANISATION OF FOREIGN TRADE IN 1983

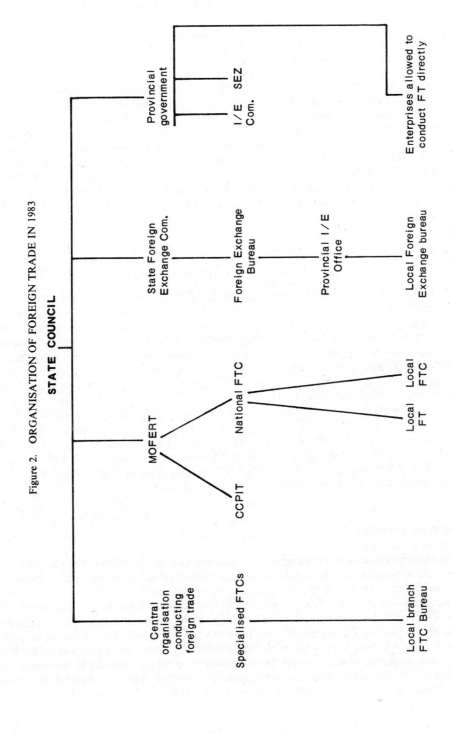

Until the "agent" system takes effect, the foreign trading system will continue to work in the following manner. If a Chinese producer – firm or state enterprise – wishes to sell a good or service abroad, or to a foreign concern in China, it must go through the FTC. The good or service is sold for a price "x1" in renminbi; this price corresponds to the domestic price of a good. The FTC then in turn sells the good or service to the foreign client for a price "x2"; this corresponds to the international price of the good in US dollars. The relationship $x = x1/x2$ is the implicit exchange rate for the transaction, and it is only very rarely equal to the official exchange rate "t" of the renminbi against the dollar. The FTC makes a profit on the exchange rate if "t" is greater than "x", and registers a loss if the contrary is true. If the transaction is profitable, the FTC transfers the profit to the parent organisation; if the transaction involves a loss, the FTC borrows from its parent organisation (usually a ministry) to pay the producer. The MOFERT has a bank account at the Bank of China, where it deposits the profits from transactions.

Imports

In the case of imports, the inverse happens. The FTC sells at x1 renminbi to the Chinese buyer a good that the FTC has paid x2 dollars to a foreign seller. The FTC therefore makes a profit if "x" = $x1/x2$ is greater than "t"; it loses money in the opposite case. Given the official exchange rate at the end of December 1980, the domestic industrial prices in China were on the average higher than the world prices for the same goods. The FTCs who imported industrial goods made profits, while those that exported Chinese produced goods lost money. In the agricultural exports and imports, the situation was the opposite; Chinese goods were lower priced than international goods.

At the national level, the MOFERT often registered considerable commercial losses: in 1981, the deficit was Rmb 6 billion[25]. There have, however, been surplus years as well[26], something that has not happened in other centrally planned economies. This type of system can only work when there is a central planning organisation that takes responsibilities for the balance of transactions at the national level; and it assumes that local entities (here FTCs and their branches) are not penalised for their losses.

The Implicit Exchange Rate

Exchange rate policies have important consequences for the volume and composition of trade at the national level. These policies also have direct effects upon the SEZs, where some degree of Chinese inputs are used in the light industrial sector.

It is very difficult to have exact information of the different exchange rates practiced in the PRC. The official rate is now quoted daily, and fluctuates in accordance with regular exchange rate practices. It is reported that the Bank of China in 1984 fixed the exchange rate on the basis of a basket of 15 currencies, including several from developing countries. In the recent past, there has been a considerable difference between the official rate and the implicit rate, granted in foreign trade transactions, to various foreign customers and clients. Table 2 summarizes this practice for several regions in 1980-1981.

As the cities of Canton, Shanghai, Beijing and Tianjin sell mostly manufactured goods and their average rate is above the average for the PRC, it appears as though the manufactured goods were more subsidized than other types. This was recently confirmed in field work in the SEZs.

A recent World Bank report[27] indicates that the implicit exchange rate for machines hovered between Rmb 2.5 to Rmb 3.0 to the dollar in 1981; it could go as high as Rmb 6 to the dollar for electronic components whereas the official rate remained at Rmb 1.9 to the dollar.

The following table also shows that the implicit exchange rate was much higher than the official rate for the year 1981, and that the renminbi was seriously overvalued for that period. The use of the implicit exchange rate allowed FTCs to reduce importations and to favour exports during that period. By 1984, American textile exporters had raised serious objections to this Chinese practice as a form of disguised subsidy.

Table 2. EXCHANGE RATES 1980-81

Region	Implicit rate (Rmb/$)	Official (Rmb/$)
Canton	3.77	1.9
Shanghai	3.25	
Beiging	3.6	
Tianjin	3.4	
National average	2.29	

Source: Guangdong *Jingji Diaocha*, n.p. 1981.

THE INTERNAL EXCHANGE RATE AND THE USE OF FOREIGN CURRENCY IN THE PRC

The different reforms in the commercial system in China over the period 1980-1984 led to the consequence that surpluses and deficits could no longer be absorbed directly by the MOFERT. At the same time, the new emphasis upon the accounting function of enterprises neither encouraged nor discouraged import or export. In order to shift these new accounting responsibilities to the local enterprise, and to promote exports over imports, the new methods were used to incite enterprises to re-organise their management structures and better control the flow of goods and services for which they were directly responsible:

    a)   the internal exchange rate and;

    b)   the possibility for enterprises to retain foreign exchange earnings.

The internal exchange rate was established by the Bank of China in January 1981 with the express permission of the State Council. This exchange rate was created to bridge the gap between the domestic Chinese prices which were subsidized directly and indirectly through collective production, and the international prices for goods. It had the effect, as well, of making imports much more expensive for the FTCs and exports much more competitive on the international market.

41

The Internal Exchange Rate

The internal exchange rate was fixed at Rmb 2.8 to the dollar and was used by all Chinese entities engaged in commercial activities. It was applied by the Bank of China and used for settlement between Chinese entities. The internal exchange rate is linked to the dollar, while the official exchange rate is linked to a basket of currencies. The official rate is used in all Chinese/foreign transactions, including tourist expenditures. There are, however, several exceptions to this rule, which will be reviewed later, most notably Sino foreign joint ventures in some of their dealings with Chinese domestic firms.

The double exchange rate had several advantages:

a) it provided an image of a non-devalued renminbi to other countries;
b) it allowed the *de facto* devaluation of the renminbi for the domestic economy;
c) it held at the same level foreign tourist exchange earnings, and transfer payments by non-residents.

The difference of approach between the pricing system of a market economy and that of the PRC means that there is almost a total incompatibility of price structures between the two systems. Thus, for instance, the internal exchange rate is probably overvalued for electronic components, and undervalued for wheat and grains. In order to convert Chinese prices into international prices, there is, practically speaking, a need for a separate schedule for each good.

The domestic situation changed greatly in 1983-1984. The official exchange rate was allowed to approach the internal rate (Rmb 2.68 in October 1984) thus providing a *de facto* devaluation of the renminbi. The central government announced an ambitious price reform for 1985-86; this will surely involve a stabilisation of the exchange rate, and probably the abolition of the internal exchange rate, at least as an officially sanctioned means of regulating transactions.

The introduction of the internal exchange rate did not, however, settle all of the foreign trade problems for the country. The table below shows that more than one-third of the exports from Guangdong province involved deficits, in spite of the use of the internal exchange rate.

Table 3. GUANGDONG EXPORTS AND EXCHANGE RATES

Implicit exchange rate (renminbi to the dollar)	No. of goods	% of exports (1979)
Less than 1.56	118	22.2
Between 1.56 and 2.8	225	40.0
More than 2.8	512	37.8

Note: 1.56 renminbi to the dollar was the official exchange rate in 1979.
2.8 renminbi to the dollar was the internal settlement rate.
Source: Guangdong *Jingji Diaocha*, n.p. 1981.

The government of the province foresaw an implicit exchange rate of Rmb 2.8 to the dollar on the average for 1981, and export subsidies of $256 million on the total of $1.6 billion exports.

Foreign Exchange Earnings

Possibilities now exist for Chinese firms to retain a percentage of their foreign exchange earnings, thus adding to the incentives to produce exportable goods. Each enterprise must have a foreign exchange account with the Bank of China; when the foreign exchange is placed into the account, a portion of it is changed automatically into renminbi at the internal exchange rate. Ministries, local governments, and certain large enterprises have the right to keep "quotas" of foreign exchange earnings in the Bank of China for future use.

These "quotas" of foreign exchange give the Chinese entity the right to use foreign exchange when the need to import specific types of equipment goods, or machine parts for production is approved. The Bank of China must always give its accord for the use of these quotas.

The "quotas" (Q1) are calculated by applying a coefficient to the surplus of exports sold during a given year when compared to a base year planned export figure (usually 1978).

The coefficients vary according to the product that is exported and according to the organisation that exports: Q1 equals q1 [X(1984)-X(1978)]. X(1984) represents exports of the producer in year 1984; X(1978) represents exports of the producer in year 1978; and X(1984)-X(1978) represents the surplus of exports between those two periods; q1 can vary from 0 per cent to 100 per cent. A great number of products necessary to the domestic economy do not earn quotas if they are exported; their q1 is 0 per cent. On the other hand, the production ministries or central organisations authorised to carry out foreign trade often have a q1 of 100 per cent. In general, this coefficient hovered around 8 to 15 per cent for 1982, a much lower rate than that of 1981. This is in part explained by the MOFERT's regaining control over the import-export functions and reducing initiatives for imports and exports by bottleneck sectors in the Chinese economy (i.e. production of energy related raw materials, for instance).

Many export oriented industries have few import needs, and therefore do not find themselves in the situation of seeking foreign exchange earnings through export. Other production enterprises which work primarily for the domestic economy may well need imports of technology and machine parts to modernize their production apparatus. The Bank of China, with the consent of the State Council, has allowed "brokers" (or "agents" as described above) to appear on the domestic market; these latter entities are intermediaries between those production units which hold foreign exchange and do not use it immediately and those that have need of for immediate import bills. These "brokers" exchange – after an authorisation by the Bank of China – foreign currency against renminbi at a maximum legal exchange rate of 10 per cent above the internal settlement rate[28].

This system gives the various production enterprises the possibility of managing their own foreign exchange accounts, and permits those with foreign currency reserves to gain an added bonus on the conversion of their currency if they exchange it with another state enterprise rather than with the Bank of China directly. In 1981, there were more than 1 500 transactions of this type, involving exchange of more than $217 million. Table 4 shows the circulation of foreign currency and indicates the changes over the period 1980-1983. For simplicity's sake the MOFERT's customs tax role and the ways in which the Bank of China and the MOFERT authorise transactions do not appear.

Towards a New Money for the SEZs

Under certain conditions, joint ventures set up in China can benefit from the internal exchange rate when they change into renminbi a part of their net receipts of foreign currency.

This measure which in fact recognises the overvaluation of the renminbi, was promulgated for two reasons:

1. To augment the foreign currency reserve of the Bank of China by preventing/discouraging the remittance of capital abroad; at the rate of Rmb 2.8 to the dollar joint ventures are encouraged to change more foreign currency than if they used the Rmb 2 to the dollar (the official rate as of 31st December 1983).
2. To diminish the imports of joint ventures into China. At the internal settlement rate, it is perhaps more advantageous for a joint venture to buy a good or a service in renminbi on the Chinese market than to import it themselves. In the case of joint ventures within the SEZs, this import substitution could accelerate the "linkage effect" on the domestic Chinese economy.

In view of regularising currency transactions in the new, more open trading system, the Central government authorised certain pilot experiments within the SEZs. The possibility of introducing a special currency for Shenzhen's SEZ was discussed in 1984; the Chinese authorities commissioned the Hokkaido Taku Bank to prepare a study on the question. The principle of this new currency was to introduce a convertible renminbi, at first to be used exclusively inside the SEZ. The initial exchange rate would be very close to the actual internal settlement rate, for reasons discussed above. For commercial reasons, the new currency would probably be pegged to the Hong Kong dollar, and not to the US dollar as the internal settlement rate is; it might also be determined as a function of a basket of currencies. The new currency, according to the study, should be controlled by the People's Bank of China, which holds important gold and foreign exchange reserves; it would be managed, however, by the Bank of China, which has much larger foreign experience.

It was announced in the Chinese press that this new type of currency would be introduced in Shenzhen in 1985, after having been approved by the State Council. Field work suggests that it will take a much longer time to establish the use of the money. According to published declarations, Chinese authorities have previewed the use of this currency as the first step towards a convertible renminbi[29]. In part, the introduction of a new single currency inside Shenzhen is designed to reduce inflation as well. By 1984, three types of currency were in use inside the SEZ: the Hong Kong dollar, the Foreign Exchange Certificate (FEC) and the renminbi. Often prices were quoted in all three involving implicit exchange rates that were variable according to the good and the location inside the zone.

This introduction of a new currency may be the solution to one of the chronic problems of the SEZ: profit remittances. Foreign partners find that their earnings within the SEZ – renminbi earnings – are not easily transferable abroad, in spite of guarantees in contracts. Under the usual circumstances, foreign firms had to buy Chinese goods with renminbi or reinvest profits in a new venture in China or the SEZs.

Inside China, economists are divided about the solutions to the problems generated by foreign currency speculation caused by the use of three types of currency. One solution put forward for the SEZs is to phase out the FEC and the reminbi currencies altogether, thus using the convertible Hong Kong dollars within the SEZs. This solution has the drawbacks of admitting the superior value of the Hong Kong currency, and at the same time linking the SEZs' monetary situation to the fluctuating fortunes of the Hong Kong dollar. It also is equivalent to extending a long-term, interest-free foreign exchange loan to the central banks of Hong Kong. There are also political problems to this solution.

A second solution would be to ban FEC and Hong Kong dollars, and use renminbi only. This solution has the advantage of extending the Chinese national currency into the zones, thus asserting full integrity of the SEZs into the country. The drawbacks, though, may outweigh these advantages: it would discourage international investment, due to the problems of exchange rates and profit remittances, it might trigger national inflation, and at the same time it would virtually open the duty free SEZ market to inland renminbi purchases allowing a substantial reduction to Chinese buyers for imported goods.

The sole use of FEC is equally unlikely, principally because these certificates are used throughout the country by tourists, and not backed by reserves, as a proper SEZ currency would have to be.

The solution of a special currency also has its supporters within the government.

Hong Kong observers, however, are sceptical about such a project for a special currency. The principal problem is linked to the narrowness of the market and the geographic limitation for the use of the currency, which means that large fluctuations would be registered in the exchange rate depending upon the transaction: trade in goods and services inside the zones, legal speculation, and even, possibly, illegal operations, such as the black market and smuggling.

These same sceptics wish to see better trained financial specialists in the SEZ before any new financial instruments are introduced.

On the whole, both Chinese and Hong Kong observers feel that the introduction of a convertible renminbi in the SEZs would benefit the SEZs as well as prepare the future for the integration of Hong Kong into the domestic Chinese economy as a Special Administrative Zone.

RECENT CHINESE TRADE PERFORMANCE

Tables 4 and 5a illustrate the foreign trade results of the open door policy inaugurated by the team of Deng Xiaoping since 1979. Thus, since 1977, the share of China in world trade has gone from 0.65 per cent to 1.13 per cent, responding to the needs for more foreign imports, and reflecting better foreign export competitivity. This performance is all the more remarkable in that it took place in a period of global trade contraction, where market shares tended to stabilize, and in spite of the drop in crude oil prices in 1981 (petroleum products represent one-fifth of the exports of the PRC). At the same time, the share of trade in the net material product grew regularly from 10.4 per cent in 1977 to 18.2 per cent before slowing down in 1982; in 1983 it reached 18.4 per cent.

Total trade, therefore, grew faster than the NMP for which the real rate of growth for the period 1977-1982 was on the average 7.7 per cent. This is an indication of the importance accorded by the Chinese authorities to the place of trade in their development strategies.

The commercial balance (see Table 5b) calculated on the basis of exports (fob)-imports (cif) shows three oscillating phases for the period 1977-1983 (see Annex 1):

1. A phase of decline between 1977 and 1979 corresponding to a series of policies directed at economic recovery and turnkey import purchases. Industrial turnkey operations and selected technological purchases rose to $6.7 billion between 1978 and 1980.

2. A recovery phase between 1980 and 1982. During this period, and notably after the cancelling of important contracts that had been signed during the previous two years, the trade balance became favourable. This contributed in a large way to the favourable foreign exchange reserve situation (more than $15 billion by 1984)

Table 4. CHINA'S TRADE AS PERCENTAGE OF WORLD TRADE

Unit: $106 million

	1973	1977	1978	1979	1980	1981	1982	1983 (January-September)
World trade (A)	1 172 450	2 289 980	2 648 371	3 331 929	4 053 474	4 013 273	3 631 133	271 325
China trade (B)	10 880	14 800	20 636	29 329	37 820	43 126	40 884	30 707
(A)/(B)	0.94	0.65	0.78	0.88	0.93	1.07	1.13	1.13

Source: *Monthly Bulletin of Statistics*, June 1984, New York, United Nations, Table 52: "World Trade by Countries and Region", pp. 106 and 126.

Table 5a. TRADE ON PERCENTAGE NMP

Unit: Rmb 100 million

	1973	1977	1978	1979	1980	1981	1982	1983
Trade (A)	220.5	272.5	355.1	454.6	563.8	717.4	756.4	860.1[a]
NMP (B)	2 318	2 644	3 010	3 350	3 038	3 940	4 247	4 673[b]
(B)/(A)	9.5	10.3	11.8	13.6	15.3	18.2	17.8	18.4

a) *China's Customs Statistics*, 1984, Hong Kong, General Administration of Customs of the PRC, June 1984.
b) Preliminary estimate disclosed in "China's Open Policy and the China Market – A Review", article published by m. Shen Mo in *China Market No. 6, 1984*, Hong Kong, Economic Information and Agency, June 1984.

Source: *Statistical Yearbook of China*, 1983, Hong, Kong, State Statistical Bureau, 1983, pp. 22 and 420 (hereafter "SSB").

Table 5*b*. COMMERCIAL BALANCE AND FOREIGN CURRENCY RESERVES FROM 1977-1983

	1977	1978	1979	1980	1981	1982	1983
Commercial balance in Rmb 100 million[1]	6.9	−19.7	−13.2	−19.0	+25	+83.6	+16.5[2]
Commercial balance in % of imports[1]	5.2	−10.5	−12.8	−6.5	+ 7.2	+24.8	+ 3.9[2]
Foreign currency reserves in $ billion[3]	2.34	1.55	2.15	2.26	4.77	11.12	14.34

Sources: 1. *Statistical Yearbook of China*, 1983, Hong Kong, State Statistical Bureau, 1983, p. 420.
2. *China's Customs Statistics*, 1984, Hong Kong, General Administration of Customs of the PRC, June 1984.
3. *International Financial Statistics*, May 1984, Washington, D.C., IMF, May 1984.

representing more than three semesters of imports, making China one of the few developing countries a net creditor on the international market. For the year 1984, the majority of experts predicted a negative trade balance.

3. China's trade volume attained a record $49.97 billion in 1984 (of which exports were $24.44 billion and imports $25.53 billion), a 22.3 per cent increase over 1983. Though the main components have not changed significantly, their relative shares did undergo dramatic movements. These movements mirror partly the domestic supply capabilities, and partly a combination of reshuffled demand structure and retarded inland transport system. On the visible and invisible account as a whole, China received in 1984 $28.89 billion in foreign exchange – or 27 per cent more than the planned target, yet spent proportionally more in foreign exchange – $29.5 billion. This expected imbalance is of course more than symbolic significance if it is China's intention to remain a net creditor nation.

But the trend continued its downward slide. In the first quarter of 1985, China suffered a $8.90 billion trade deficit as imports jumped 54.4 per cent to $6.04 billion, whereas exports rose by only 2.7 per cent to $5.15 billion, when compared with the same period last year. The items loading the import bills were consumer durables such as colour televisions, washing machines, etc., (100 per cent), followed by machinery electrical instruments (90 per cent) and so on. In view of these events, it is likely that China might incur a larger trade imbalance for 1985 than previously, unless effective measures are adopted to curb this trend.

The Structure of Trade in 1983

Table 6 indicates that the patterns of trade for the PRC in 1983 showed a large degree of complementarity between exports and imports; this is a typical developing country trade structure which adopts an inter-branch international division of labour as opposed to an intra-branch one, characteristic of developed countries trading among themselves. The textile sector (with the exception of synthetic fibres) represented 23.6 per cent of PRC exports; the energy sector represented 21.0 per cent with oil exports representing 19 per cent alone. Among the other large export sectors are agricultural products (fruit and vegetables, and meats exported principally towards Hong Kong). The southern provinces (and especially the Island of Hainan) export tropical products such as tea, coffee and cacao.

Imports were largely in the area of heavy industrial goods, (see Table 7) and in particular transport goods and machine and equipment goods (which represent 20 per cent of Chinese

Table 6. STRUCTURE OF CHINESE TRADE IN 1983

	Exports		Imports	
	1983	1983	1983	1983
	%	Mil. Rmb	%	Mil. Rmb
SO + S1* = N1	13.3	58.32	14.8	62.48
N2	9.0	39.38	11.8	49.87
N3	21.0	92.02	0.5	2.19
Primary production	43.3	189.72	27.2	114.54
N4	5.6	24.67	14.9	62.77
N5	5.2	22.69	25.1	105.87
N6	5.5	24.06	18.6	78.65
Heavy industry	16.3	71.42	58.6	247.29
N7	23.6	103.61	2.8	11.90
N8	7.9	34.82	5.1	21.66
Light industry	31.5	138.43	7.9	33.56
N9	8.9	38.75	6.4	26.43
Total	100.0	438.32	100.0	421.82

Note: The nomenclature used here is derived from the SITC (1) Rev. II with the following operation: N1 = S0 + S1.
* S0 + S1 = N1 Food, beverages.
  S2 + S4 = N2 Crude materials, except fuel.
  S3 = N3 Energy.
  S5 = N4 Chemical products.
  S6 = (61, 62, 63, 64, 65) = N5 Metallic and non-metallic mineral manufactures.
  S7 = N6 Machinery and transport equipment (61+65+84+85) = N7 Textile, garment and leather.
  S8 = (84, 85) + (62, 63, 64) = N8 Other light industrial goods.
  S9 = N9 Commodities not elsewhere classified.
Source: China Customs Statistics, 1984, General Administration of Customs of the PRC, Hong Kong, June 1984, pp. 20-23.

Table 7. EVOLUTION OF THE STRUCTURE OF TRADE 1981-1983

%

	Exports			Imports		
	1981[1]	1982[2]	1983[3]	1981[1]	1982[1]	1983[2]
Primary products	46.6	45.0	43.3	36.5	39.6	27.2
Food and beverages	13.6	13.4	13.3	17.4	22.5	14.6
Crude materials	9.3	7.8	9.0	18.7	16.1	12.1
Energy	23.7	23.8	20.9	0.4	1.0	0.5
Manufactured goods	53.4	55.5	56.7	63.5	60.4	72.8
Chemical products	6.1	5.4	5.6	11.9	15.2	14.9
Light products and manufactures of metal[3]	21.4	19.2	19.6	18.3	20.3	29.3
Machinery and transport equipment	4.9	5.7	5.5	26.7	16.6	18.6
Other manufactured goods[4]	16.5	16.6	17.1	2.5	2.5	3.7
Commodities not classified elsewhere	4.1	8.1	8.8	4.1	5.8	6.3

Source: 1. State Statistical Bureau, 1983.
2. China Customs Statistics, 1984.
Note: 3. Leather, rubber, paper, textile products, metallic and non-metallic manufactures.
4. Travel goods, footwear, garment, scientific instruments.

imports); chemical products, and more especially fertilizers and organic chemical products, which represent close to 15 per cent of imports. This situation will change as vertical integration takes place in China with respect to the oil sector and the chemical sector. The reserve ratio of Chinese chemical exports has however decreased from 51 per cent in 1981 to 40 per cent in 1983.

The road towards modernisation that remains ahead for the Chinese authorities, and the choices this implies for the development of light and heavy industry, is reflected in the statistics on ferrous metal products and non-ferrous metal product imports. These two categories represented more than Rmb 10 billion in 1983, or about one quarter of total Chinese imports. After the slow-down on large projects such as the Bao Shan steel works, it is difficult to see how the Chinese will be able to reduce these imports in the near future.

The Evolution of the Trade in the PRC

Since 1981 exports have had a constant evolution characterised by a shift from raw materials to manufactured goods. This is due first to the effect of prices on petrol products. In the period 1981-1984, there was a tendency for prices to fall in the energy sector. However, the moderate decentralisation of trade in the country also allowed production units selling manufactured goods to export more freely, thus earning more foreign currency for their own needed imports. The trade frictions that occurred between the United States and the PRC in 1983 do not seem to have had an impact upon the aggregate level of trade, as the share of light industrial exports and metals grew between 1982 and 1983. The taxonomy used by Chinese authorities does not permit a close examination of the evolution of textile exports, which represent about one-quarter of total PRC exports. As Table 8 shows, these exports are divided almost equally between capital intensive textile products and garment exports which are labour intensive. PRC imports of textile products are decreasing. These imports are chiefly raw cotton, wool and synthetic and artificial fibre materials.

Table 8. CHINA'S IMPORTS AND EXPORTS OF TEXTILE PRODUCTS

renminbi million

	Exports			Imports		
	1981	1982	1983	1981	1982	1983
Raw materials (silk, cashmere, cotton)	3.61	6.42	5.52	25.77	13.50	6.78
Textiles	26.42	37.36	35.25	30.04	22.24	14.06
Clothing, footwear	24.86	29.63	33.67	–	–	–

Source: Compiled by OECD Development Centre from:
  – *China's Customs Statistics*, 1984;
  – *State Statistical Bureau*, 1983.

The evolution of the structure of imports is quite different from that of exports. The share of imports taken by raw materials, and most particularly by foodstuffs, is directly related to the annual harvest yields in the PRC. Any deficit in domestic production must be met with cereal imports, usually from the United States. In the manufactured goods sector, there has been a continual drop. The equipment goods component of this sector shows a particularly irregular pattern (26.7 per cent of imports in 1981, 16.6 per cent of imports in 1982, 18.6 per

Table 9. CHINA'S IMPORTS OF HIGH TECHNOLOGY PRODUCTS

$ million

	1979	1980	1981	1982	1983
Imports of equipment goods (world total) (A)	3 950[1]	5 376[1]	5 104[1]	2 793[2]	3 533[2]
High technology imports (total OECD) (B)	667[3]	1 079[3]	869[3]	584[3]	901[3]
(B)/(A) (%)	25.5	16.9	20.1	17.0	20.9

Sources: 1. World Bank.
2. China's Customs Statistics. The data are extrapolated from a series in renminbi; data converted into US dollars at the official rate of exchange for the given year.
3. OECD.

cent in 1983). This is an echo of the new policy formulations during 1982, when central government authorities took measures to avoid overinvestment in heavy industry. To this tendency to import heavy equipment goods is associated a similar trend to purchase high technology goods. Table 9 traces the evolution of Chinese imports of high technology over the period 1979-1983.

The evolution of the relationship of equipment goods imports expressed in dollars and imports of high technology reveals the preference of Chinese authorities for acquiring selected items of technology rather than turnkey projects. The rate of technological importation reached its highest point in 1982 and 1983, as Table 9 shows.

Other large imports in 1983 were cereals (11.5 per cent), paper and wood (4.1 per cent) and textile fibres (3.9 per cent).

The Terms of trade

Table 10 gives the evolution between 1978 and 1982 of the price index for imports and exports. From this can be deduced the evolution of the terms of trade.

Table 10. TERMS OF TRADE

	1978	1979	1980	1981	1982[1]
Import price index	100	119.4	139.2	145.5	131.0
Export price index	100	113.7	131.3	134.9	128.1
Terms of trade index	100	95.2	94.3	92.7	97.8

1. Estimates.
Source: MOFERT.

The rapid growth of exports is due, in part (about 30 per cent), to the effect of prices on exports over the period 1978-1982. The rate of average annual growth of the price of exports was 6.4 per cent, whereas the value of exports grew on the average of 20.8 per cent for the same period (annual rate cf. Table 9). In the same fashion, the effect of prices explains about 40 per cent of the growth of imports, reflecting most probably the emphasis placed on the purchase of equipment goods and turnkey operations for the period 1978-1981. For the latter, the structure of offer is more oligopolistic, and the relative prices are higher than those of finished goods. Overall, the terms of trade worsened between 1978 and 1981, before bettering in 1982. This reflects the domestic policies of the central government on the whole, rather than international conditions.

The Direction of Trade

In the taxonomies proposed in the Tables 11 and 12, it appears clear that Japan was the first commercial partner of China in 1983 if Hong Kong is excluded. Japan buys 20.4 per cent of China's exports, and has a 25.9 per cent market share of China's imports. It is followed by Hong Kong with 26.2 per cent and 8.0 per cent respectively. Developing countries in Asia and the Middle East are in third position (21.0 per cent and 6.8 per cent respectively), followed by the EEC (11.0 per cent and 15.1 per cent). The United States bought 7.7 per cent of PRC exports and furnished 12.9 per cent of its imports. Although trade with Eastern Bloc countries has been small, recent signs of "rapprochement" with the Soviet Union in particular may mean larger exchanges between the PRC and the COMECON countries[30].

Chinese imports come overwhelming from OECD area countries. The market share of OECD area countries attained 74.5 per cent in 1981. In 1983, after trade frictions with the United States on textile exports, and after a favourable harvest in China, the US market share dropped from 22.7 per cent to 12.9 per cent; at the same time, the market shares of Japan and the EEC grew respectively from 20.6 per cent to 25.9 per cent and from 11.0 per cent to 15.1 per cent. But neither Japan nor the EEC was able to surpass previous high water marks (28.5 per cent in 1978 for Japan and 21.3 per cent in 1979 for the EEC). Recent measures announced by the US customs authorities have effectively changed the rules concerning the attribution of quotas to countries of origin where products have been successively transformed

Table 11. GEOGRAPHIC SHARE OF EXPORTS 1978-1983

%

	1978[1]	1979[1]	1980[1]	1981[2]	1982[3]	1983[3]
Japan	17.6	20.2	22.1	22.1	21.9	20.4
United States	2.8	4.3	5.4	7.0	8.0	7.7
EEC	12.3	12.6	12.6	11.2	9.5	11.0
Other OECD	5.2	5.7	4.5	3.7	3.4	3.1
Total OECD	37.9	42.8	44.6	43.6	42.8	42.2
USSR	2.4	1.8	1.2	0.6	0.7	1.4
Other East Europe	8.9	8.5	5.8	3.2	3.0	3.0
Total East Europe	11.3	10.3	7.0	3.8	3.7	4.4
Hong Kong	25.9	23.7	23.8	24.5	23.7	26.2
Other developing Asia			17.0	20.0	21.3	21.0
Other developing countries	24.9	23.2	6.8	8.1	8.5	6.2
Total developing countries	50.8	46.9	48.4	52.6	53.5	53.4
Total world	100	100	100	100	100	100

Source: 1. World Bank.
  2. State Statistical Bureau, 1983.
  3. China's Customs Statistics.

Table 12. GEOGRAPHIC SHARE OF IMPORTS 1979-1983

%

	1978	1979	1980	1981	1982	1983
Japan	28.5	25.2	26.4	28.6	20.6	25.9
United States	6.6	11.8	19.5	21.6	22.7	12.9
EEC	19.1	21.3	14.1	12.2	11.0	15.1
Other EEC	19.2	14.5	13.4	12.1	14.6	14.4
Total OECD	73.4	72.8	73.4	74.5	68.9	68.3
USSR	1.9	1.6	1.4	0.7	1.3	2.1
Other East Europe	7.8	8.4	7.6	3.7	4.8	4.2
Total East Europe	9.7	10.0	9.0	4.4	6.1	6.3
Hong Kong	0.7	1.4	2.9	5.8	6.9	8.0
Other developing Asia	16.2	15.8	8.5	6.9	9.2	6.8
Other developing countries			6.2	8.4	8.9	10.9
Total developing countries	16.9	17.2	17.6	21.1	25.0	25.4
Total world	100	100	100	100	100	100

Source: 1. World Bank, 1983.
2. *State Statistical Bureau*, 1983.
3. *China's Customs Statistics*, II, 1984.

Table 13. COMPOSITION OF CHINESE IMPORTS
FROM THREE MAJOR OECD PARTNERS IN 1983

%

	Japan	United States	German Federal Republic
Food products	0.3	24.3	0.3
Raw materials	2.0	16.8	0.2
Total: primary commodities	2.3	40.1	0.5
Chemical products	9.4	23.4	22.7
Processed raw materials and metals/ non metals	45.9	3.7	31.7
Equipment, foods	30.2	20.7	39.7
Total: heavy industry	85.5	47.8	94.1
Textiles/clothing/leather	3.6	0.9	1.1
Other light industrial products	6.3	10.4	3.9
Total: light industry	9.9	11.3	5.0
Other products	2.3	0.8	0.4
Total	100	100	100
(Rmb million)	10 904	5 463	2 397

Source: *China Customs Statistics*, II, 1984.

in intermediate countries. The large exporting countries such as China, Hong Kong and South Korea have felt the effects of these measures most severely. Enacted on 7th September 1984, the measures were aimed at reducing US textile imports. These measures will surely have an effect on Sino-United States trade in 1985; previously, similar non-tariff measures enacted by the United States against the PRC cost the United States $2 billion in cereal trade with China.

Table 13 gives the composition by product group of Chinese imports coming from Japan, the United States, and the Federal Republic of Germany. An important contrast appears between the FRG and to a lesser extent Japan, whose exports to the PRC are made up principally of heavy industrial products, and the United States, which exports agricultural and wood products to the PRC. It should be recalled, however, that "other light industrial products" include scientific equipment and control instruments, for which the United States is the principal supplier for the PRC.

Aside from the United States, other OECD area countries export the majority of their heavy industrial goods towards the PRC. These exports allow some of the exporting countries to shore up domestic industries that are declining (chemicals, steel, mechanical construction machines).

The geographic distribution of Chinese exports presents the same type of evolution as that of imports: there is a slight decline of market share for OECD area countries and a small growth of market shares for COMECON countries. The principal difference is in the role of Hong Kong, which buys – irrespective of the year – a quarter of Chinese exports, and thus gives the impression of a high level of "South-South" trade between China and developing countries in 1983 (53.4 per cent of exports were directed towards these countries). The share of PRC exports absorbed by OECD area countries was 68.3 per cent. Japan remains the principal OECD area client with 20.4 per cent in 1983. The.EEC absorbed 11.0 per cent and the United States absorbed 7.7 per cent for the same period. But these figures are somewhat distorted due to the role of "entrepôt" trade through Hong Kong.

Hong Kong's Entrepôt Trade

Re-export statistics from Hong Kong do not allow for a detailed understanding of the geographic destination of these re-exports, except for a small number of countries. For these countries, Table 14 shows the impact of Hong Kong entrepôt trade on their economies. Real exports from the PRC to a country "a" are the sum of the exports of the PRC to "a" plus the sum of PRC exports to Hong Kong which are then re-exported to "a". The same holds true for imports. Re-export of PRC goods through Hong Kong represented $2 523 million or 11.5 per cent of PRC exports. Re-export of goods to the PRC from Hong Kong represented, in 1983, 7.4 per cent of PRC imports. On the whole, it is as if Hong Kong were acting as a large trading company which took charge of one-tenth of PRC trade, and which kept for itself another one-tenth of that trade.

The market shares of the principal trading partners of the PRC take into account Hong Kong entrepôt trade. It is important to note that both Japanese and the US market shares have been on the rise, whereas the share of Hong Kong is declining as can be seen in Table 15. This entrepôt trade has political advantages as well. The PRC conducts hidden trade with Taiwan and South Korea through Hong Kong. The PRC also uses the entrepôt function to re-import goods that it has exported; thus, it profits from the port and warehouse facilities in Hong Kong to re-ship goods to the north or south of China itself. This "repurchase" of PRC exports by China amounted to $127 million in 1983, or about 5 per cent of the total goods re-exported

Table 14. HONG KONG'S ROLE IN CHINA TRADE (1983)

$ billion

	Chinese exports	Hong Kong re-exports	Real exports	Chinese imports	Hong Kong re-exports	Real imports	Real balance
United States	1 698	706	2 404	2 731	189	2 920	−516
Japan	4 481	142	4 623	5 452	434	5 886	−1 263
Taiwan + Republic of Korea	0	197	197	0	199	199	−2
China	0	127	127	0	127	127	0
Indonesia	96	209	305	149	31	180	121
Singapore	563	152	715	113	33	146	569
Hong Kong	5 151	−2 523	3 228	1 696	−1 562	134	3 094
Rest of the world	9 336	1 117	10 453	10 950	676	11 626	−1 173
Total	21 916	127	22 043	21 091	127	21 218	825

Source: China's Customs Statistics (II, 1984).
China's Trade Report, June 1984.
Hong Kong Review of Overseas Trade in 1983.

Table 15. GEOGRAPHICAL DISTRIBUTION OF REAL IMPORTS/EXPORTS FOR CHINA IN 1983

%

	Exports	Real exports	Imports	Real imports
United States	7.8	11.0	13.0	13.9
Japan	20.4	21.1	25.9	27.9
Taiwan + Republic of Korea	0	0.9	0	1.0
Indonesia	0.4	1.4	0.7	0.8
Singapore	2.6	3.3	0.5	0.7
Hong Kong	26.2	14.6	8.0	0.6
Rest of the world	42.6	47.7	51.9	55.1
Total	100	100	100	100

Source: Calculated by OECD Development Centre from China's:
– *China's Customs Statistics*, 1984 (1);
– *China Trade Report*, June 1984;
– *Hong Kong Review of Overseas Trade in 1983;*
– *Census and Statistics Department*, Hong Kong, 1984.

towards the PRC. This is one of the many unusual roles that Hong Kong fills for the PRC. The balance of trade for the PRC and the regions defined in the taxonomy established above is represented in Table 16. On the whole, the trade surplus with developing countries allows the PRC to cover the structural deficit with OECD area countries.

It is difficult to assess the recent evolution of PRC trade with the rest of the world. It appears, however, that PRC-Hong Kong trade provides a growing surplus for China and this is confirmed in the preliminary data for the first trimester of 1984, where the trade surplus with the territory was HK$14.4 billion against HK$9.6 billion for the same period in 1983[31]. It is also important to recall that the PRC trade balance appeared to become more favourable in 1984, as it produced a surplus of Rmb 503 million in the first trimester of 1984[32]. However, by the end of 1984, there was a sharp drop in the terms of trade, and the PRC was again registering important trade deficits ($1.7 billion beyond 1984).

54

Table 16. TRADE BALANCE BY REGIONS 1981-1983

Renminbi million

	1981[1]	1982[1]	1983[2]	Real balance 1983[3]
Japan	−2 377	1 716	−1 942	−2 526
United States	5 368	−4 774	−2 064	−1 032
EEC	−377	−58	−1 541	n.a.
Other OECD	−3 127	−3 818	−4 495	n.a.
Total OECD	−11 251	−6 881	−10 045	n.a.
USSR	−51	−189	−241	n.a.
Other East Europe	−161	−464	−466	n.a.
Total East Europe	−212	−653	−707	
Hong Kong	6 876	7 319	8 109	6 188
Other developing Asia	4 807	5 492	6 313	n.a.
Other developing countries	−208	387	−2 019	n.a.
Total developing countries	11 475	13 198	12 403	n.a.
Total world	−12	5 663	1 650	1 650

Source: 1. *State Statistical Bureau*, 1983.
2. *China's Customs Statistics.*
3. Cf. Table 12.

The real trade balance can be defined as the difference between real exports and real imports; this eliminates the effect of re-export trade through Hong Kong. It becomes clear, then, that the "real" trade deficit with the United States is reduced by one-half, whereas the "real" trade deficit with Japan grows even larger. At the same time, the surplus registered with Hong Kong is reduced considerably.

The Balance of Payments

The evolution of the trade balance presented in the preceding section has contributed for a number of years to the surplus in the current account of the PRC balance of payments. Other items have had the same effect, such as remittances, tourism, etc. As the flow of capital into the PRC has been larger than the flow outward in the period 1979-1984, the foreign exchange reserves have grown in a healthy fashion. Table 17 illustrates this evolution of the balance of payments between 1979 and 1983.

If the balance of goods and services showed a spectacularly favourable evolution, the current accounts payments were not as favourable (in spite of the increase of tourist receipts); this was due in part to the rising cost of international transportation and the rising costs for licensing fees and other royalties paid out in connection with technology transfers. At the same time, remittances by overseas Chinese seem to have stabilized.

At the level of capital flows, the table shows some interesting trends for the PRC:

1. the appearance of important direct investment flows, as well as international loans both bilateral and multilateral;
2. the drop in middle- and long-term debt flows;

55

3. the increase in 1981 of short-term loans by the PRC to LDCs, due in part to the high foreign exchange reserves.

Tables 17 and 18 illustrate the evolution of the external debt and the foreign exchange reserve of the PRC since 1979.

The figures concerning the external debt represent the sum of credits accorded by the IMF and the World Bank, foreign governments, Japanese energy credits; they exclude liabilities, deferred payments and private sector bank loans.

Table 17.   BALANCE OF PAYMENTS PRC

$ billion

	1983	1981	1979
Exports fob	23.3	22.0	15.0
Imports fob	19.8	22.7	17.1
Trade balance	3.5	−0.7	−2.1
Invisibles (net)	1.8	2.7	0.5
of which: Remittances	(0.6)	(0.5)	(0.7)
Current balance	5.3	2.0	−1.6
Direct investment	0.4	0.3	–
Non-bank borrowing	0.2	0.6	+0.2
Bank borrowing	0.1	−0.6	+2.4
Change in net reserves	+6.0	+2.3	−0.6

Source: 1. *Financial Times,* 29th October 1984.
2. *World Bank* 1983, p. 133.

Table 18.   CHINA'S EXTERNAL DEBT 1979-1982

$ million

Total 1979-1982 (1983 in parentheses)	No. of agree-ments	Amount agreed (pledged)	Used	Remarks
Borrowings	61	15 214 (1 050)	10 871	
1. Cash through the Bank of China	22	7 560	7 560	Mainly for 22 key projects including the Baoshan steel complex and for petroleum investment. All repaid for 1979-82.
(Suppliers' credits)		(20 000)		Mainly from Great Britain, France, Italy, Canada, Sweden, Australia, Belgium, Norway, Argentina, West Germany and Denmark.
2. Loans from foreign governments	29	5 233	1 478	Long and medium-term loans from Japan, Belgium, Denmark, Kuwait, Italy and others.
3. World Bank and IMF	10	1 791	935	
4. Others		898	898	Borrowed by local governments and units of the ministries on their own account, including suppliers' credit of 268 million.

Source: *China Quarterly,* No. 100, December 1984, pp. 826-827.

56

The figures reflect at the same time – with a small gap – the evolution of the foreign exchange reserves of the PRC. In the period 1979-1984, the PRC became a net creditor. However, certain Chinese officials have estimated that the development and exploitation of offshore oil reserves, the construction of a nuclear power station and the cost of upgrading the internal transport system will lead the PRC to borrow more than $20 billion from 1985-1990, and a total of $50 billion from 1985 to 1995. The situation of foreign exchange reserves in 1984 allows the PRC to credibly present itself as a small "risk" for international loans, and with a triple "A" credit rating, to obtain loans at low rates. However, there is also the possibility that the high foreign exchange reserves will provide another argument to potential multilateral lenders: why lend at 3 per cent to a country that is capable of placing its own sizable foreign exchange reserves on the international monetary market at 17 per cent? Some might think it best to reserve preferential loans to developing countries with much less favourable foreign exchange reserves.

Chinese authorities have foreseen this argument, and have pointed out that this favourable foreign exchange reserve is a temporary situation, and more especially, that the amounts of foreign capital necessary for the modernisation of China are far greater than present reserves. Some private forecasting institutions foresee a large trade deficit for the PRC by 1988, due in part to equipment goods imported by the country. This trade deficit is estimated to be $9.5 billion (by 1988); there is also a possibility that direct foreign investment will grow steadily, and at the same time the country will contract a larger private sector bank debt, receive larger international loans and credit packages in the future.

The Chinese authorities have kept carefully in mind the debt crisis precipitated in Latin America, and closer to home, in a country like South Korea. The policy of diversifying sources of finance for development, and negotiating them carefully with partners, is meant to lessen the political and economic risks involved. This logic is at the source of the decision of the PRC authorities to liberalise investment codes for the country, and to create special environments, like the SEZs and the 14 coastal cities, where foreign direct investment will be encouraged but carefully monitored at the same time.

The Financial System and External Transactions

China trade is largely financed by foreign banks outside China, foreign state banks which provide buyers and sellers credits, the Bank of China, the Investment Bank of China, the China International Trust and Investment Corporation and by private capital. Representative offices of foreign banks operating in China are generally not permitted to engage in profit making activities.

The capacity of Chinese enterprises and accounting units to finance trade with foreign entities is affected by their access to foreign exchange. By 1985, local governments and enterprises had easier access to foreign exchange than before 1979. Domestic enterprises are now allowed to retain a portion of their profits rather than being required to turn all profits over to the government. These enterprises now retain a portion of their foreign exchange earnings in special accounts. Thus export enterprises have an external source of foreign exchange.

Local governments also retain a portion of the foreign exchange they earn through taxes on export enterprises giving them an internal source of foreign exchange. Branches of the Bank of China are also allowed to lend a portion of their foreign exchange earnings to domestic enterprises or local governments. The local investment and trust corporation is likewise able to provide foreign exchange through its business dealings for the locality. The local import-export commission can allocate a certain amount of foreign exchange and can seek additional amounts through the MOFERT. Finally, localities receive foreign exchange allocation amounts through the central government and local government economic plans.

In the period 1983-1984, the financial system of the PRC was greatly modified[33]. The People's Bank of China became the central bank, and its domestic and international activities were given over to other banks specially created for this purpose (see Figure 3). These latter were previously departments of the bank.

Figure 3. CHINA'S DOMESTIC BANKING SYSTEM

Decision-taking Authority	*Secondary authority*
Ministry of Finance Est. 1949 (MOF)	*People's Bank of China* Est. 1948; began operating as Central Bank, 1984 (PBOC)
People's Construction Bank of China (PCBC) Est. under MOF, 1954 Indep. from PBOC, 1984	*Industrial and Commercial Bank of China* (ICBC)
China Investment Bank Est. 1981 (CIB)	*Bank of China* (BOC) Est. 1908
	China Intl Trust and Investment Corp. (CITIC) Est. 1979
	Agricultural Bank of China (ABC) Est. 1979
	People's Insurance Company of China (PICC) Spun off from PBC 1984

Source: China Business Review, March April 1985, p. 18.

By 1984, there were seven major banks in China, a national insurance company and the China Investment and Trust Corporation involved in financial affairs.

The People's Bank of China (PBOC) has the following functions:

- to formulate monetary rules and regulations for China;
- to issue currency and to control its circulation;
- to exercise control over credit and deposit interest rates;
- to control foreign exchange and precious metals;
- to examine and to approve the establishment, merger and dissolution of monetary .organisations (CITIC, PICC);
- to oversee operations of specialised banks (ICBC, BOC, ABC);
- to operate on behalf of China in the international monetary system;
- to act as a state treasury.

The People's Bank of China has its headquarters in Beijing, and has branches and sub-branches throughout China.

The Bank of China (BOC) is responsible for dealing with the foreign exchange accounts of the country. The Bank of China:

- handles foreign exchange transactions;
- holds deposits in foreign currencies and in renminbi related to foreign exchange operations;

- grants import and export loans;
- grants foreign exchange loans;
- handles international settlements in connection with trade and non-trade transactions;
- buys and sells foreign exchange reserves;
- handles all overseas remittances;
- concludes agreements with foreign governments and central banks;
- provides guarantees to Chinese enterprises in business transactions;
- is moving into the international syndicated loan market and the Euro-Bond markets.

The Bank of China has branches throughout China, and branches in New York, London, Singapore, Luxembourg, Karachi, Manchester, Paris, Tokyo.

The State General Administration of Foreign Exchange Control is charged with fixing and announcing the exchange rates, and at the same time formulating all foreign exchange decrees.

The Investment Bank of China (IBC) specialises in raising foreign funds for investment in China. Generally, it seeks medium- and long-term credit from international monetary institutions. So far, the main task of this bank has been to serve as an intermediary in handling World Bank loans.

The Industrial and Commercial Bank of China (ICBOC) was founded on 1st January 1984 to take over the commercial role of the People's Bank. It administers deposits for industrial and commercial enterprises, and grants industrial and commercial loans for production and commodity circulation in addition to making short-term equipment loans. This bank also receives savings deposits in towns and cities, and issues, when approved, commercial and home-buyer loans. Recently, the ICBOC has been allowed to deal in foreign currency in Shenzhen, thus breaching the monopoly of the Bank of China in this area.

The People's Construction Bank of China (PCBC) specialises in handling state allocations for construction and in making loans for capital construction. It may also provide loans to Chinese-foreign joint ventures for capital construction.

The Agricultural Bank of China (ABC) extends credit for agricultural projects and it receives rural deposits. It exercises a leadership role over the credit co-operatives as well. These latter are collectively owned banking institutions in rural areas which take deposits and provide loans.

The China International Trust and Investment Corporation (CITIC), described below, plays a key role in financing development projects. It raises funds on international capital markets, as well as engaging in trust investment and financial leasing.

The People's Insurance Company (PIC) handles domestic and foreign reinsurance; no foreign concern is yet allowed to offer insurance services in the PRC.

China's activity on the international financial markets is steadily increasing. The China International Trust and Investment Corporation floated a Yen 10 billion twelve-year bond with a coupon of 8.7 per cent and the Fujian Province Investment and Enterprise Corporation issued a Yen 5 billion ten-year bond with an 8.55 coupon[34]. The Bank of China was scheduled to float a Yen bond issue on the Tokyo Samurai market in December 1984.

Shenzhen itself has experimented with the sale of stock on the domestic financial market; these stocks can be purchased by other Chinese enterprises[35]. There has likewise been discussion of re-opening the Shanghai Stock Exchange[36].

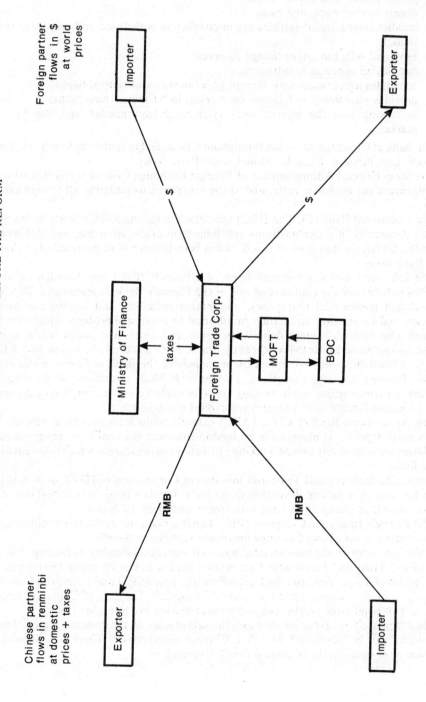

Figure 4. FINANCIAL TRANSACTIONS BEFORE THE REFORM

Foreign partner
flows in $
at world
prices

Importer

Exporter

$

$

Ministry of Finance

taxes

Foreign Trade Corp.

MOFT

BOC

RMB

RMB

Chinese partner
flows in renminbi
at domestic
prices + taxes

Exporter

Importer

60

International Transactions

Before the reform of the financial institutions was carried out in the early 1980s, there was no contact between the Chinese importers or exporters and foreign firms. Chinese enterprises had to use the intermediary of the Foreign Trading Corporations for all transactions abroad. The FTCs handled negotiations and commercial transactions, as well as a number of monetary transactions. Figure 4 presents financial flows within and without China, and highlights one of the major problems that has plagued China: the disjunction between domestic prices and international prices.

Domestic Prices

One of the principles of central planning in China is to minimize the role of prices in the allocation of resources. The overall equilibrium of the economy is accomplished by a system of rationing of goods. Prices have a determining importance on the measure of return on profits of an enterprise and on the size of fiscal revenues. The structure of relative prices affects the standard of living of rural households in respect to urban households, and it has a determining influence on consumption choices.

Prices can be fixed at the national, provincial or local level, depending upon the range of distribution of a given article or service. The different price bureaux depend hierarchically upon the central bureau of prices, which in turn is under the Plan. The central bureau co-ordinates price structures proposed by other ministries. In some cases (notably for oil prices), the State Council must give its accord as well.

Prices are generally determined by costs. These costs are principally the enterprises profit margin, remittance taxes and operating costs taken together. Agricultural prices are generally determined by the necessity of maintaining a pegged price between agricultural products and industrial products.

A certain number of products can have prices that are deliberately set below the costs of production, either because they are necessities (wheat, rice, etc) or because they are industrial inputs that are subsidized (i.e. electricity).

Domestic prices are therefore disconnected from world prices, and the equilibrium between world offer and demand intervenes rarely in the fixing of domestic prices. One notable exception are goods that cannot be produced in China and that are imported (mostly capital goods, and high technology). In these cases, the domestic price is the international price, converted into renminbi at the official exchange rate, with import taxes added.

INSTITUTIONS HANDLING FOREIGN DIRECT INVESTMENT

Figure 5 presents the outline of Chinese institutions that handle foreign investment. The distinction between trading activities and investment activities is sometimes unclear, it is not surprising that some of the institutions described as handling trading activities are also cited here.

Among the most important institutions that are specifically investment related are:

The Foreign Investment Control Commission (FICC), which was created in 1979 at the same time as the Import-Export Commission. Originally, the FICC depended directly on the

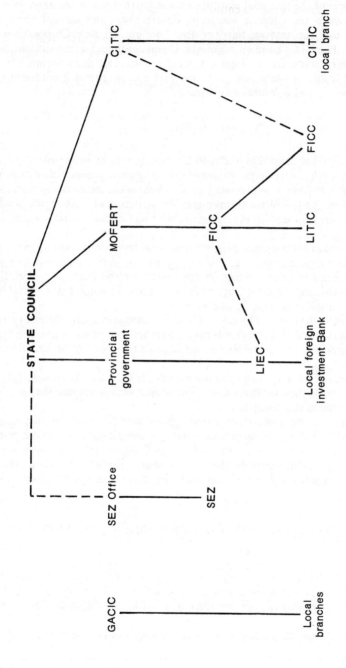

Figure 5.   INSTITUTIONS IN CHARGE OF FOREIGN INVESTMENT IN 1983

—— primary management authority
‑ ‑ ‑ consultative role

State Council, supervised directly by Gu-Mu, the Vice-Premier Minister responsible for the SEZs. The FICC has the following mandate:

1. Examine and approve joint ventures projects;
2. Research and formulate laws, decrees and regulations;
3. Organise, consider, and conclude foreign economic co-operation agreements;
4. Investigate and research the general state of international development and trade[37].

In fact, the real role of the FICC is to co-ordinate and to approve contracts between Chinese and foreign partners. In a centralised economy, this is a vital function. It is at this level that investment is generally tabulated (number of contracts approved). The FICC has the power to approve or refuse any FDI contract above $3 million. However, the SEZs, the newly designated 14 coastal cities, the municipalities of Shanghai, Beijing and Tianjin and the provinces of Fujian and Guangdong are not subject to the FICC, but rather have their own investment approval bodies. The municipality of Shanghai for instance, can approve contracts up to $30 million; the SEZs have no real upper limit on investment. Elsewhere in China, any investment, regardless of the amount, must be approved by the FICC if it is in the form of a contractual joint venture or an equity joint venture.

The China International Trust and Investment Corporation (CITIC) was set up in October, 1979. Like the FICC, it too reports directly to the State Council. The CITIC acts as an agent for Chinese partners: it contacts potential investors outside China, arranges co-financing for joint ventures, takes care of acquiring advanced technology and specialised capital goods. The CITIC is specifically mentioned in the Law on Joint Ventures as the body for arranging Equity Joint Ventures. It has the authority to set up financial institutions abroad, possible joint ventures, for the purpose of raising funds for lending outside China. It has local branch offices. The CITIC can also use its own fund to become a partner in a joint venture.

The Local Investment and Trust Commissions (LITC)[38] have been created in the provinces and the municipalities that have the most contact with foreign investors. The functions are very close to those of the CITIC, and they often act in concert with that body. The LITCs depend hierarchically upon the Local Import/Export Commission (LIEC), mentioned above in the section on trade. This latter is a provincial level authority with ties to the MOFERT. The LIEC are aware of all local initiatives for FDI, and when the amount in question is larger than their authority to approve, they transmit the request to the FICC. Local Foreign Exchange Control Offices, whose role is not clearly defined, seem to work closely with the LIEC to control the foreign exchange account.

Foreign Direct Investment in the People's Republic of China

As part of the new economic climate created by the Chinese authorities after 1978, the government decided that for the first time since the withdrawal of the Soviet technical assistance at the end of the 1950s, new direct foreign investment in the PRC would be allowed. This was clearly to be part of the new "open door" policy of China[39]. The Vice-Chairman of the National Planning Commission, Yuan Baohua, confirmed this new orientation in a public announcement in 1978. The decision to solicit foreign direct investment in China was a pragmatic one taken by the Deng team. The immense capital needed to finance the modernisation of the country could scarcely be raised domestically; to use existing foreign aid resources was unrealistic, as China had not sufficiently established its place in the various multilateral aid agencies; the government did not want to become dependent upon bilateral aid sources either; international loans were discussed, and China moved cautiously into the area as a borrower, principally from the Government of Japan. The third source of external capital was foreign direct investment. This latter would, however, involve a substantial shift in domestic policy, as it would involve a greater foreign presence in China as well as a more transparent account of the economic workings of the country. New legal codes would also have to be developed, to attract foreign investors and assure them of sufficient protection against expropriation or radical shifts back to Maoist policies. This opening to foreign direct investment was justified as in keeping with the Leninist injunction to use external capital for socialist development[40].

The principal obstacle from the foreign investor's point of view was the general lack of experience on the part of the Chinese authorities in dealing with the international trade and finance systems in place. There was also suspicion that the political upheavals of the recent past would repeat themselves, and foreign companies might find themselves expropriated or expelled should a new wave of xenophobia spread through the country. The Chinese quickly understood that the PRC would have to produce legislation protecting the introduction of foreign investment in the country. Japanese officials provided the Chinese with a compendium of legislation on foreign investment enacted in different developing countries in which the Japanese trading companies were operating. The Japanese efforts at providing suggestions on legislation came up against a serious problem: the lack of civil legal structures within the PRC itself. The foreign investment legislation was to be more precise than any domestic law, because it would have to adapt itself to legal categories such as private property, individual rights, corporation rights, contract law and patent law. This is, of course, an on-going process, and the literature on the subject is now large[41].

The "Law of the People's Republic of China on Joint Venture Using Chinese and Foreign Investment" was finally approved on 1st July 1979 by the second session of the Fifth National People's Congress[42]. Although the law was general in its language, it left leeway for special interpretation to provide the necessary guarantees to the foreign investor. A State Foreign Investment Commission was created to oversee the investment process, and auxiliary agencies sprang up to aid the implementation of the law (such as the China International Investment Trust Corporation, or the Provincial Investment Enterprise Corporations which have come into existence to promote investment among Hong Kong, Macao and overseas Chinese). The 15 articles of the law were soon complemented with other laws, including the law regulating the creation of the SEZs. Explanations followed from Beijing, as investors found areas of the law that needed clarification[43].

The idea of using foreign capital to promote the modernisation of the PRC was contested by a certain number of interest groups in the country, and opposition had to be constantly checked and mollified when the authorities made new moves towards opening the country.

The direct investment law is the groundwork for the SEZ legislation that followed. In this sense, the SEZs may be seen as particular cases of foreign investment legislation. This special status helped them get off the ground in the first two years of operation (1980-81), when the rest of the Chinese economy was undergoing a "readjustment". One of the more surprising effects of the quest for foreign investment was the growing autonomy of the provincial governments in their dealings with foreigners. This autonomy caused a certain amount of disruption in the capital investment programme throughout the country, and the central government began to limit the powers of the various provincial authorities in initiating foreign investment for capital construction projects.

64

The search for foreign investment sources led the Chinese to be more specific about the taxonomy of economic co-operation. The terms used, even in Chinese, tend to be fluid; the following terminology is the most commonly used by the Chinese authorities: (Chinese pinyin transcription underneath)[44].

Equity joint venture (He zi jingying)	– A limited liability corporation in which Chinese and foreign partners invest jointly in and operate a corporation, share the profits, losses and the risks.
Contractual joint venture (co-production) (Hezuo jingying)	– Often also called a co-production or co-operative project. It can involve the foreign partner providing the technology, and a capital share. Foreign investors are repaid on a schedule of return negotiated beforehand. The Chinese partner usually provides land, materials, the work force, basic buildings, services, etc.
Compensation trade (Buchang maoyi)	– The foreign partner provides technology and equipment. The Chinese partner repays with goods produced by the same equipment. The foreign partner markets the goods internationally.
Processing/assembly (Lailiao jiaogong or Laijian zhwangpei)	– Foreign partner supplies raw materials or intermediate goods to Chinese partner for assembly or manufacture, according to foreign design. Chinese are paid for their services, and the foreign partner markets the product abroad.
100 per cent foreign owned (Du zi chi ye)	– The foreign partner covers all production costs, labour, and utility costs. Enterprise is simply located within China.

The equity joint venture formula has particular attractions for the Chinese:
  a) it divides risks and profits according to equity shares;
  b) it usually involves technology transfer agreements;
  c) it usually places the marketing of the joint venture's product in the hands of the more experienced foreign partners;
  d) it often involves real capital transfers into China.

It is therefore significant that this type of agreement has not initially had the success the Chinese had hoped for. Table 19 indicates that by the end of 1982, of the pledged capital investment for the whole of the PRC, less than 6 per cent was joint equity venture. This reluctance on the part of foreign investors to set up equity joint ventures may have been due to the rather vague legal code, which many felt did not adequately specify the conditions under which joint ventures could be dissolved. The trend in FDI, however, appears to be in favour of joint ventures (contractual and equity). By mid-1985, official sources had announced that 1984 alone more than 741 joint venture agreements had been signed; this figure is three times the number signed between 1979-1984. By April 1985, the number of joint ventures – equity and contractual – had risen to 930 for all China.

For the period 1979-1983, the most favoured form of investment was the co-operative or co-production agreement (also called "contractual joint venture"). One of the attractions in the co-production was the way in which the contracts are negotiated: virtually everything was left to the discretion of the two partners, to be spelled out in the final written contract. This type of arrangement did not pass through the normal chain of administrative approval. It has

Table 19. BREAKDOWN OF DIRECT FOREIGN INVESTMENT IN CHINA

Table 19. BREAKDOWN OF DIRECT FOREIGN INVESTMENT IN CHINA

Unit: $ million

Item	1979-1982			1979-1983			Jan. 1979-June 1984		
	N	Kp	Kr	N	Kp	Kr	N	Kp	Kr
J.V.	83	141	n.a.	190	321	n.a.	362	331	n.a.
Co-production	792	2 726	n.a.	1 047	2 950	n.a.	1 372	3 500	n.a.
Offshore oil	12	999	n.a.	23	2 041	n.a.	31	2 400	n.a.
Other	905	1 092	n.a.	998	1 400	n.a.	1 137	800	n.a.
Total	1 853	4 958	n.a.	2 300	6 711	2 600	2 900	7 231	3 300

Kp = Capital pledged
Kr = Capital realised
Source: Intertrade, October 1983; JETRO China Newsletter, No. 51; Official MOFERT announcements, Hong Kong, November 1984.

been reported that many Chinese partners have initiated co-operative agreements under the guise of co-production agreements, in order to have them approved at the central level, where administrative machinery is well disposed to these agreements; they are then changed into joint venture agreements, once the administrative go-ahead has been given.

The term of "co-operative agreement" covers a great number of contractual arrangements, and is used loosely in most statistical literature. In this type of agreement, profit shares are negotiated beforehand, and do not necessarily correspond to the equity shares that were contributed under the original contract. Once the contractual period is over, the production unit normally reverts to Chinese control, thus ensuring at least the transfer of machine equipment, although not necessarily the design or ingredient element that makes a product attractive on the international market[45]. The principal advantage from the Chinese point of view of this type of contract is the fact that the foreign partner makes most of the capital investment and takes most of the production and marketing risks. It has also been pointed out that co-production contracts are favoured because they are not included in the State Plan, and therefore escape quotas and close surveillance. It can also mean, however, that the co-production enterprises do not receive adequate supplies, because they are not provided for in the State Plan (this is a problem even for those enterprises that are included in the Plan at times). There is also an added problem for foreign investors in estimating the political influence that their Chinese partners may have, something that could become crucial should the State supply system break down and the production unit be obliged to import quality raw materials from abroad.

Compensation trade involves a transfer of technology from the foreign investor in return for finished products. The foreign partner once again takes all the risks, and once again the equipment becomes Chinese property after the expiration of the lease. Normally, all goods must be marketed outside the PRC, but in the case of an enterprise that has paid all foreign currency bills, goods may be sold on the domestic market, if they are already on the import list[46].

Table 19 presents the relevant data for China's use of foreign funds since the beginning of the open door policy. It is interesting to note that compensation agreements, co-production agreements, offshore oil exploration agreements and 100 per cent foreign owned firms (represented by "others"; most of these are located within the SEZs) represent the overwhelming majority of foreign investment agreements, both in number of agreements and in value. This is the case principally because these agreements are much easier to implement, once official permission is secured; they can also involve modest investment sums and in fact are often little more than disguised trade agreements[47]. The financing of co-production units is also the worry of the foreign partner. Although foreigners may have an easier time securing international finance, they must, in the end deal with the Bank of China – the only foreign exchange Bank in China – which does not follow western practices in guarantees of performance and repayment needed for loans elsewhere. The complex relationships between foreign firms and financing units with the Chinese structures make it unlikely that a local foreign enterprise will undertake a request for financing.

Investment problems

The failure of the PRC to help foreign investors secure loans or facilitate other forms of credit during the period 1979-83 caused a slowdown in foreign direct investment throughout the country. The response of PRC authorities was firm: foreign exchange loans are expensive to finance. Chen Muhua, Minister of Foreign Economic Relations, at first expressed doubts over the use of commercial loans. She said that once China began earning enough foreign exchange from the export-oriented industries, then commercial Chinese bank loans might be possible. The Bank of China preferred to lend available foreign exchange on the international market where it could reap 17-18 per cent interest rates to creditworthy clients[48]; such interest rates are well beyond the profit-earning capacity of most PRC enterprises. By December 1984, however, the Bank of China's SEZ branches – especially the Shenzhen branch – were actively lending both renminbi and foreign currency to joint venture enterprises set up in the zones. This policy could account for the increase of joint ventures, within Shenzhen, as well as Xiamen, where the Bank of China is a member of the development company consortium[49].

Related to this problem of foreign exchange loans to Chinese enterprises is investment in equipment. During the period 1980-1983, the heavy industries of the PRC were in a slump, and equipment goods and high technology imports were drastically reduced. This was partly the result of protectionist measures on the part of ministries that were attempting to sell their own Chinese-made products on the domestic market. It was also part of the national policy to reduce the external debt of the country by reducing imports of all kinds. The government, with the policies of "readjustment", was anxious to reduce the unco-ordinated planning at the enterprise level, and therefore reduced the initiative capacity in many cases to import capital equipment goods.

With characteristic prudence, the central government is therefore trying to complement commercial loans and borrowing schemes with foreign direct investment as a source of foreign currency and a principal tool of modernisation. The expectation level for foreign investment is high indeed, considering the relatively unprepared situation in the PRC[50].

Recently, officials of the PRC have made known certain modifications that are planned for the Joint Venture Law. The guideline for these modifications was presented in a recent seminar held in Hong Kong[51]. This text represents a gloss on the legislation passed in September 1983 on joint ventures[52]. The principal innovations were a clearer definition of "legal person" (a Chinese and foreign legal person would now have the same statute), the extension of the tax *holiday* from three to five years (the first and second profitable year will be tax exempt, the remaining three, 50 per cent tax relief), and certain types of industries would be allowed a larger tax holiday (those involved in low-profit operations such as farming and forestry or those joint ventures established in remote areas; this tax holiday can be made retroactive).

For the first time, the question of the access to the domestic market was openly discussed. The primary object of most joint ventures in the PRC is to promote exports, and this for two reasons. First, exports earn foreign currency for the joint venture, and therefore provide profits that can be taken outside the PRC by the foreign partner, and at the same time provide the Chinese partner with its foreign exchange. The joint venture can also buy any international products it needs, pay foreign staff workers, etc. The second reason is that firms must maintain a foreign currency balance if they wish to have access to the domestic market. It is a form of quality control. The products sold on the international markets will be tested in a very competitive environment (an admission that the domestic market does not have the same quality standards) and thus provide the joint venture with an objective standard for production, which in turn can up-grade goods sold on a domestic market.

However, the domestic market is – and has been – only made accessible to products that are normally imported into China[53]. The policy of access to the domestic market is not entirely clear. It is stated that products "urgently needed by China's modernisation" or those "already imported from abroad" can enter the domestic market under certain circumstances and at certain ratios of total joint venture production. This would indicate a policy of import substitution, something the Chinese have to date refused to endorse. Even if there is access to the domestic market, it is unlikely that foreign currency could/would be used to pay for the product. With earnings in renminbi, it is unlikely that a firm would want to massively invest in the domestic market, unless an attendant reform of foreign exchange laws is introduced, allowing joint ventures to convert renminbi into foreign currency and remit it abroad.

Joint ventures, which do not have a place in the State plan, were not always able to ensure a fluid supply of raw materials nor ensure a marketing procedure for their goods. As a result, Chinese authorities have made it clear that "in principle" joint ventures should be treated on an equal basis with State enterprises in securing supplies and in fixing prices for goods on the domestic market. In the latter situation, the joint venture will have to include its production in the State plan quotas, so that adequate measures can be taken to ensure distribution.

The supply side of the joint venture economy is a critical issue for Chinese planners, as it involves a price differential of considerable magnitude. Domestic prices for raw materials and semi-manufactured goods are much lower than on the international market, and it is not possible for the State planning authorities to give priority to joint ventures that are exporting goods abroad. The price of raw materials is therefore fixed at international standards, or slightly lower; only in some cases, it is equivalent to the domestic price[54]. This complicates the problem of re-introducing goods into the domestic market, as the price differential for the raw materials would have to be recuperated in the selling price, something that is not possible at the moment, given the fixed domestic price for goods based upon domestic input prices.

Remittance of after-taxed income and salaries of staff in the joint venture was another thorny problem. The initial agreement by the Chinese authorities was to allow up to 50 per cent of after-tax income to be remitted abroad in foreign currency; joint venture enterprises found that this was not enough. The domestic market of China did not offer sufficient commodities to entice workers (especially those from Hong Kong and Macao) to spend within the domestic economy. The new 1983 regulations allow individuals to apply for remittance of a larger proportion of their salaries; they can "apply to the Bank of China for permission to remit outside China all the remaining foreign exchange after deduction of their expenses in China".

The wage structure of joint ventures was set in the initial agreements according to Chinese wishes to avoid discrimination against Chinese staff. The principle of equal pay for equal work was instituted. However, in practice, there were problems. By Chinese law, senior positions must be shared by Chinese and foreign staff. This meant that senior Chinese staff were paid the same salaries as those of senior foreign staff regardless of the qualifications of the former. The 1983 regulation implicitly admitted that the Chinese were deficient in business experience, and therefore salaries could be decided upon in light of actual work performance rather than rank[55].

The legislation concerning disputes became much clearer as a result of the new regulations. It allows for arbitration either in China or in another country, a step towards accepting international arbitration. Agreements for mutual protection of investment and investors have also been permitted.

The SEZs and Foreign Direct Investment

The SEZs were designed to attract foreign direct investment in a much larger fashion than the rest of China. Not only were they accorded a special legal status; the usual administrative structures were simplified, and new agencies of the state banking, insurance, labour and planning administrations were set up. The SEZs were to be a "special case" of foreign investment, enclaves, where foreigners could reside and enjoy many of the advantages of the world beyond the Chinese borders. The Chinese themselves were determined to make the four zones models of development practice. The incentive packages are in the vanguard of changes within China, and as such herald many of the concessions granted later to firms within China.

The 14 Coastal Cities and Beyond

In April 1984, the Chinese authorities announced that 14 coastal cities would be opened to further preferential foreign direct investment. It has been reported that Deng Xioaping himself was active in – and perhaps even behind – this decision[56]. The 14 coastal cities were to have economic and technical zones to develop newly emerging technology-intensive industries. In November 1984, a symposium was held in Hong Kong to promote the new open cities, and specific preferential treatment was spelled out for foreign investors[57]. The seminar turned out to be a very successful event. By the end of the two week period set aside for investment pledges, more than $2.2 billion had been pledged; $2.5 billion worth of agreements were the object of negotiation; this raised the total pledged foreign investment for China to almost $10 billion by end-1984[58]. There have been official statements suggesting that the entire coastal area of China could be opened soon to foreign direct investment, something that would involve nine provinces and almost half the population of the country[59].

In this regard, the present study of the SEZs intends to clarify the performance record of the first areas opened to foreign direct investment, and attempts to set the stage for discussing the role of foreign direct investment in the modernisation of China, now so clearly the path upon which the country is set.

The SEZs, and since 1984, the 14 coastal cities, are the demonstration that cycles of continuity and rupture are the means by which China has opened to the international trading and finance systems. The importance of these areas for China's trade will depend a great deal on the development of transport infrastructure for shipping and streamlined trade and customs administrations; these conditions are now being met in several of the large coastal cities and all four of the SEZs. These latter have, in fact, served as laboratories for many of the trade and investment reforms introduced into China after 1980. The role of the SEZs in refining new economic policies for China has been crucial. So too has been their success in selling Chinese reforms to the international community. What follows is the analysis of that bold development experience.

NOTES AND REFERENCES

1. See for instance the three-volume study of the Chinese economy published by the World Bank, *op. cit.* Numerous articles have appeared in Chinese and English in specialised journals. No. 100 (December 1984) of *China Quarterly* was devoted entirely to the readjustment period.

2. See in particular Christopher Howe's analysis of the period 1979-1983 in his article "China's International Trade : Policy and Organizational Change and their Place in the Economic Readjustment", in *China Quarterly*, No. 100, December 1984, p. 813 ff. Several useful journalistic articles have also appeared in *Intertrade* (Hong Kong), October 1983, April, May, August 1984, January 1985.

3. "If we include commissioned borrowings, the Bank of China has extended on its own account total foreign exchange loans amounting to $19.7 billion for the period 1979-1983. These include $870 million for major state construction projects, $690 million for energy and transport projects (sponsored by some 40 industrial departments) and $397 million for technological renovation in locally controlled medium- and small-scale enterprises". Howe, C., *op. cit.*

4. Among the many good books on the subject, see the excellent presentation of economic history of the imperial period presented as a prolegomena to present economic conditions in China in M. Elvin, *Patterns of the Chinese Past*, Stanford, 1973.

5. Wang Gungwu treats the fluctuating degrees in which the Chinese felt this superiority in his contribution to *The Chinese World Order*, J.K. Fairbank (Ed), Harvard, 1968. See also in the same volume Lien-sheng Yang's contribution on the Sinocentric world order.

6. See Chapters 1-2 of J.K. Fairbank's *Trade and Diplomacy on the China Coast: The Opening of the Treaty Ports*, Stanford, 1974.

7. See Fairbank, *op. cit.* p. 28.

8. See M. Mancall, "The Ch'ing Tribute System: An Interpretive Essay", in *The Chinese World Order, op. cit.* For a representative document of the period on the economic functions of the tribute system see the Qing compilation: *Huang ging zhi gong tu* (Illustrations of the Regular Tributaries of the Imperial Qing), Palace Edition, 1761, Tong Gao et alia.

9. Fairbank, *op. cit.*, p. 31. "Outsiders could have contact with China only on China's terms, which were, in effect that the outsider should acknowledge and enter into the Chinese scheme of things and to that extent become innocuous ... Tribute was the first step towards sinicizing the barbarian and so neutralizing him."

10. For the classic study of the Canton trade see H.B. Morse, *International Relations of the Chinese Empire*, London, Watson and Viney, 1930, and S.F. Wright, *Hart and the Chinese Customs*, Belfast, 1950.

11. Matthieu Ricci, Nicolas Trigault, *Histoire de l'expédition chrétienne au Royaume de la Chine*, Paris, 1618, p. 894.

12. The famous Ming expeditions of Cheng Ho (1403-1433) were in some measure trade missions. They helped establish junk routes which Europeans then took over in the 16th century.

13. See for instance I. Hsu, *The Rise of Modern China*, New York, 1970 and his *China's Entrance into the Family of Nations : The Diplomatic Phase*, 1856-1880, Harvard, 1960, as well as the classic of H. Morse, *International Relations of the Chinese Empire, op. cit.*, and especially Fairbank, *op. cit.*

14. See Fairbank, *op. cit.,* p. 47 ff.

15. Fairbank, *op. cit.* p. 60 ff. From 1775 to 1795, the Company derived more than one-third of its revenue from the country trade with China.

16. Fairbank, *op. cit.,* p. 74.

17. Chang Chuh-tung, Governor of Guangdong (1885-1889) Hunan (1889-1894) and Hupei (1896-1906), was but one of the famous neo-conservatives who espoused the cause of self-sufficient development, something that the Chinese have taken up in present times as "autocentric" development.

18. For a recent treatment of this question, see the account given by Jonathan Spence in his *The Gate of Heavenly Peace: The Chinese and Their Revolution, 1895-1980,* New York, Viking Press, 1981.

19. See the analysis in Epstein, *China's Economic Development: The Interplay of Scarcity and Ideology,* Ann Arbor, 1975, p. 124 ff. And: Frances Moulder, *Japan, China and the Modern World Economy,* Cambridge, United Kingdom, 1977.

20. Epstein, *op. cit.,* p. 127. Epstein says that what modernisation did occur, took place in the industrial and commercial sectors of the economy, and precisely because of the import of foreign techniques. Agriculture, the base of the economy, remained labour intensive and traditional.

21. See M. Elvin, *Patterns of the Chinese Past,* Stanford, 1973 and F. Moulder, *China, Japan and the Modern World Economy,* Cambridge, UP, 1978.

22. S. Ellis, "Decentralization of China's Foreign Trade Structures", in *Georgia Journal of International and Comparative Law,* vol. 11:2, 1981.

23. Xu Dixin *et alia.* "Socialist Modernisation and the Pattern of Foreign Trade" part D, *China's Search for Economic Growth,* Beijing, 1982.

24. For a review of these reforms, see the article by Wang Dacheng "Reforming the Foreign Trade Structure" in *Beijing Review,* Vol. 27, No. 43, 1985, pp. 4-5. Wang notes: "The current reform will first separate government administration from enterprise management ... Enterprises will engage in foreign trade under the state plan, conduct business accounting and assume sole responsibility for profits and losses within the framework of state policies. The new foreign trade enterprises will act as agents in China's import and export business. Under this system, the enterprises provide services, are entrusted to handle import and export business and receive a service charge, while production enterprises are responsible for losses and gains ... In the past, all exports were handled by foreign trade corporations. They bought goods from production enterprises and took responsibility for gains and losses ... The agent system, under which the production enterprises assume sole responsibility for their gains and losses, will help solve these problems. For the description of the new FTCs, see *Guide To China's Foreign Economic Relations and Trade: Import-Export,* Hong Kong 1984.

25. *Euromoney,* October 1983 "Why the renminbi must be devalued" by L. Goostad.

26. V. Falkenheim, "Chinese Trade Policy" (mimeo: Paper presented to Conference on Emerging Pacific Community Concept", 24th-26th October 1983, Georgetown University.)

27. *China's Socialist Economy, op. cit.,* 1983.

28. This varies considerably throughout China, as field work has demonstrated. It can go as high as 80 per cent above the internal settlement rate in extreme cases, these latter usually occur in Guangdong Province. For a review of the specific problems of currency reform within Shenzhen, see Chen Wenhong, "Problems of New Shenzhen Currency", *Guangjiao jing* (Wide Angle), Hong Kong, May 1985, pp. 40-46, (in Chinese).

29. *China Business Review* 7th August 1984; for a critical review of this currency problem, see Chen Wenhong, "What are Shenzhen's Problems", *Guangjiao jing,* (Wide Angle), Hong Hong, No. 149, pp. 40-45, April 1985.

30. In December 1984, Ivan Arkipov, a Soviet Deputy Prime Minister officially visited the PRC. Three agreements were signed: a technical and economic co-operation agreement, a science and technology agreement and an overall economic co-operation agreement. Promises were made for an agreement in early 1985 for more trade between the two countries during their respective Five Year Plans 1986-1990. This may increase co-operation among the two planning agencies of the PRC and the USSR. Sino-Soviet trade was projected to rise to $1.42 billion for 1985.

31. *China Trade Report*, August 1984.

32. *China's Customs Statistics*, 1984.

33. See D. Brotman "Reforming The Domestic Banking System", *China Business Review*, March-April 1985, p. 17 ff.

34. "China's Yen for Bonds" *Far Eastern Economic Review*, 5th July 1984.

35. "The First Stock-issuing Experience", *Intertrade*, May 1984.

36. "Shanghai Bank to Issue Shares for Enterprises" *Ta Kung Pao*, 23rd August 1984.

37. S. Ellis "Decentralization of China's Foreign Trade Structures", in *Georgia Journal of International and Comparative Law*, vol. 11, February 1982.

38. The opening of the PRC is the subject of a great deal of recent literature. It is concisely chronicled in Chapter 9, of *China A Business Guide*, Tokyo, Japan External Trade Organisation, 1979. See also the excellent presentation of the recent policy reforms in an economic context in S. Ho and R. Huenemann, *China's Open Door Policy: The Quest for Foreign Technology and Capital*, Vancouver, University of British Colombia Press, 1984. The Soviet Union had a number of joint ventures with the PRC during the 1950s in Xinjiang Province, but these were terminated in 1960.

39. The opening of the PRC to foreign investment is chronicled in Chapter 9, of *China, A Business Guide*, Japan External Trade Organisation (Tokyo, 1979). The PRC had joint ventures with the Soviet Union in the Xinjiang province, but these were terminated in 1960.

40. Li Qiang, a senior economic advisor, stated that "from a Marxist-Leninist point of view, the methods we are adopting for utilising foreign capital in joint ventures to develop China's resources are correct in principle." *Beijing Review*, 27th April 1979.

41. For an overview of the problem, see T. Gelatt, "Doing Business with China: The Developing Legal Framework", *The China Business Review*, November-December 1981, pp. 52-56; J. Cohen, "Equity Joint Ventures: 10 Pitfalls that Every Company Should Know About", *The China Business Review*, November-December 1982, pp. 23-30. J. Stepanek, "Direct Foreign Investment in China", *The China Business Review*, September-October 1982, pp. 20 ff. Papers were presented to the UNIDO conference for the Promotion of Industrial Investments in China (Guangzhou 7th-11th June 1982) on the legal aspects of foreign investment (by Chinese authors). One of the most comprehensive publications to date is the *Guide to China's Foreign Economic Relations and Trade* (Investment Special), Hong Kong 1983 (in English and Chinese). This latter gives full text for recent laws relating to foreign investment in the PRC. See also *Annex 1*, "China's Legal System".

42. The text of the law is included in the *Guide to China's Foreign Trade, op. cit.*, p. 206, ff.

43. See *JETRO China Newsletter*, No. 44, p. 143 ff.

44. For more details, see Christopher Howe, Y.Y. Kueh. "China's International Trade: Policy and Organisational Change and Their Place in the 'Economic Readjustment'" *China Quarterly*, No. 100, December 1984, p. 813 ff. See also, *Zhongwai hezi jingying giye* (Chinese-Foreign Shared Capital Enterprises) by Wang Yihe *et alia*, Shanghai, Shanghai shehui kexueyuan, April 1984.

45. The Coca Cola and Pepsi Cola operations in Shanghai and Shenzhen are good examples of this. With the 1985 patent law, licensing and processing technologies are protected.

46. See S. Shirk and J. Stepanek "China's Five Year Reform Programme", *China Business Review,* November-December 1983, pp. 8-10. More recently, other concessions have been made to foreign partners for access to the domestic market.

47. See the discussion of this point in Howe and Kueh, *op. cit.,* p. 813 ff.

48. Stepanek, *op. cit.,* p. 24: "In August 1981, the Bank of China co-managed two syndicated loans to a Spanish highway corporation for $45 million; in September, it co-managed and managed syndicated loans totalling $140 million to two Brazilian corporations .... as recently as June, it participated in a $175 million syndication to Denmark's privately funded Eximbank. These loans do not include the bank's outstanding loans to the eurocurrency market."

49. It was reported that the Bank of China intended to make domestic loans to the sum of $2 billion before the end of 1985 to medium- and small-sized enterprises in the country. There has already been a significant impact on productivity as a result of domestic foreign currency loans: "According to statistics of 611 projects that had paid off their loans (totalling $535 million) by the end of 1981, output value of the enterprises involved had increased by Rmb 5 700 million; the foreign export volume increased by Rmb 2 600 million, and the supply of commodities on the domestic market rose by Rmb 3 100 million during the loan period. That is to say, one US dollar of loan increased the output value by Rmb 10.65 and the export value of Rmb 4.07 during the loan period". Chao Yongyi, "Expanded China's International Banking Business", *China Market,* April 1983.

50. In late 1982, it was calculated that the PRC was hoping for more than $4 billion in direct investment (1 001 provincial level projects and 121 national level). Stepanek, *op. cit.,* p. 27. More recently at the investment symposium seminar held in Hong Kong in November 1984 more than 190 agreements were initialled or signed representing a total of $2.50 billion – a sum far above expected FDI for the 14 coastal cities.

51. Presented in paper by Yu Yimin, Vice Director of Foreign Economic Relations of the PRC, at a seminar organised in November 1983, in Hong Kong, entitled Seminar on China's Joint Venture Law. Paper title: "Creating More Favourable Investment Environment for Foreign Investors". See also a presentation of the new regulations in Masao Sakurai, "Investment-related Laws and Regulations", *JETRO China Newsletter,* No. 48.

52. On 20th September 1983, the State Council of the PRC promulgated the Regulations for the Implementation of the Law of the PRC on Joint Ventures Using Chinese and Foreign Investment. The first joint venture law dates from July 1979. See Chapter 6 for a thorough discussion of Chinese legal system.

53. "For instance, the export proportion for the China Schindler Elevators Company Ltd. one of the early established joint ventures in 1980, it is stipulated that 20 per cent of the production will be sold in China". Yu Yimin, *op. cit.,* p. 9. This is really an import substitution joint venture from the beginning and as such, an unusual situation. The Volkswagen Shanghai assembly plant signed in October 1984 was also largely an import substitution agreement.

54. According to article 65 of the joint venture law, 6 raw materials – gold, silver, platinum, petroleum, coal and timber – must be sold at international prices to joint venture enterprises. The law goes on to state "when purchasing export or import commodities, the suppliers and buyers shall negotiate the price, with reference to the prices on the international markets, and foreign currency shall be paid. Except for the cases mentioned above, the prices for purchasing all other materials on the Chinese domestic market shall correspond with state-set prices and be paid in renminbi". Services such as transportation, advertisement, water, gas, electricity and heat can also be paid in renminbi by the joint venture.

55. Articles 93 and 94 of the law stipulate that "the salary and bonus systems of joint ventures shall be in accordance with the principle of distribution to each according to his work, and more pay for more work". Salaries can now be fixed by vote of the board of directors "in light of actual conditions".

56. See the report in *Beijing Information,* January 1985.

57. The comparison of investment incentives is made in Annex 3. The fourteen coastal cities include: Beihai, Dalian, Fuzhou, Guangzhou, Lianyungang, Lantong, Ningbo, Qingdao, Qinhuangdao, Shanghai, Tianjin, Wenzhou, Yantai, Zhangjiang; Hainan island is included as well in the new "open areas". See also "Preferential Treatment for Investors in Open Cities", Xinhua, in English, as reported in *Summary of World Broadcasts* (SWB) FE/7796/C/1, 9th November 1984.

58. Ma Meili "Promoting the Open Policy in Style", *Intertrade,* December 1984.

59. Gu Mu himself made statements to this effect in December 1984. By mid-1985 this had not taken place. However, it was announced that the Yangtse and Pearl River Deltas, and the southern part of Fujian province, including Xiamen, Zhanzhou and Quanzhou would be opened. "Xinhua" News Release as reported in *Summary of World Broadcasts* (SWB) FE/W1379/A/9, 13th March 1985.

Chapter 3

PROMOTING FOREIGN TRADE: EXPORT PROCESSING ZONES IN THE ASIAN CONTEXT

The first Export Processing Zone (EPZ) created in Asia was located in Taiwan (established in 1966)[1]. There were political as well as economic reasons for this decision[2]. Later, at the end of the 1960s, other Zones were created throughout Asia. This experience had particular significance for the Chinese, as it constituted a precedent for the creation of the Special Economic Zones (SEZs) in Guangdong and Fujian provinces.

There is evidence that the Chinese authorities took a careful look at the performance of the other EPZs in Asia when deciding the conditions for the SEZs[3]. The Chinese choose the term "special economic zone" *(jinji tequ)* to differentiate their effort from the existing export processing zones in Asia, although the conditions pertaining in the SEZs were to be very similar to those in EPZs in general.

The rise of export processing zones and free trade zones is a well chronicled chapter in development economics and several recent studies have researched the performance of the numerous zones throughout the world; there are now some 400 "free zones" around the globe,

Table 20. TYPES OF FREE TRADE ZONES

Type	Characteristics
Free trade zones	Geographically limited, usually near port; free trade with rest of world. Merchandise may be moved in and out, stored, opened or inspection and repackaged as needed;
Export processing	Also called "industrial free zones"; zones created to motor export of assembled goods and light manufacturing goods. No customs on goods moving in and out of EPZ. Tax advantages; regulatory incentives; land utility costs low;
Free ports	Infrastructure related to free trade zone or export processing zone which creates duty free shops, hotels, casinos, housing. No duty on goods sold in shops;
Enterprise zones	Domestic zones in which special conditions are offered to businesses in order to create new outlets in distressed areas. Primarily oriented towards indigenous businesses;
Free banking zones	Also called "offshore" banking facility. A relaxation of foreign currency controls, interest rate ceilings on deposits and reserve requirements, attracts non-resident foreign currency dominated business. May be functional or mailbox.

although the number of free trade zones and export processing zones is much smaller[4]. The distinction drawn elsewhere concerning the type of economic zone is useful to recall (see Table 20).

The evolution of the free trade zone towards the export processing zone is a modern phenomenon. The first free trade zones date from the 18th century; they represented trade arrangements that allowed for repackaging, storage and re-export of goods. Free trade zones were located along the established international trade routes, and were mostly in the hands of trading companies.

The idea of a zone with special economic regulations for commercial purposes gave way to a concept of a zone with industrial manufacturing bases, linked to the commercial activities of the free trade zone. The phenomenal growth of these zones was a testimony to the success of the idea: by 1980 more than 55 zones were in operation in 30 developing countries; 35 more were being studied or planned (20 different countries).

The Chinese decision to establish the SEZs should be seen in the context of this development. One of the general goals of the EPZ was to aid domestic import-dependent industries to enlarge their industrial base, principally with foreign capital. Another was to give export industries, both domestically owned and joint ventures, a more competitive advantage by lowering the tax rate. Imports of raw materials, intermediate products, equipment and machinery are not subject to duty. The goods produced in the zones are normally destined for export onto the international market.

Traditionally, EPZs have been created in Asian developing countries to promote exports; the more specific goals in creating the EPZs were important precedents to Chinese expectations. They include:[5]

- use of foreign investment to upgrade technology and indigenous industrial base;
- foreign exchange earnings;
- creation of local employment;
- upgrading management and labour skills;
- creation of some links to the domestic economy.

Aside from these expectations on the part of the host country, the advantages offered to the foreign investors in the EPZs included special administrative arrangements for lower customs duties (or no duties at all), and tax exemptions. These conditions often also apply to export industries located outside the EPZ, but the added advantage of duty-free importation and the relative concentration of export industries with the zone made them more attractive than the export production unit located in isolation within the host country.

Table 21 presents basic data for selected Asian EPZs.

The Chinese Experience

The Chinese perception of advantages to the foreign investor are somewhat different, and merit closer examination; beyond cost and labour considerations for foreigners, the Chinese authorities felt they were offering attractive "intangible" benefits. They considered the following as attractions for investors:

1. Chinese authorities offered a guarantee of political stability for the economic development of the SEZs (in contrast perhaps with the domestic economic situation);
2. There was a greater latitude of economic operations inside the SEZs (once again, in reference to the domestic economy);

Table 21. BASIC DATA ON EXPORT PROCESSING ZONES
IN SELECTED ASIAN COUNTRIES (AROUND 1980)

Country Location/EPZ	Year of establishment	Area (acres)			Premium (1000 $/acre)	Rent ($/acre/year)	Lease terms (years)
		Planned	Developed	Allocated			
Indonesia							
Batam Island[a]	1978	9 143[b]	2 224[c]		103	629[d]	30 [20]
Jakarta							
Tanjung Priok	1973	82[e]	82			255 224 /470 448[f]	30 [20]
Malaysia							
Johore							
Pasir Gudang	g	g	g	g	40-50	92-184[h]	60
Senai	1977	100	90	67	27-40	92-184[h]	60
Melaka							
Batu Berendam	1973	52	53	28	24	808	99
Tanjong Kling	1973	170	170	71	17-20	81-276	99
Penang							
Bayan Lepas	1971	304	263	160	40-50	400	60
Prai	1972	416	303	212	35	400	60
Prai Wharf	1972	42	38	15	i	i	60
Pulau Jerajak	1972	406	20	20	j	400	60
Selangor							
Ampang Ulu Klang	1973	50	50	43	27	810	99
Sungai Way Subang	1972	141	141	104	40	1 200	99
Telok Panglima Garang	1974	49	49	40	27	810	99
Philippines							
Central Luzon							
Bataan	1972	853	855	597		7 982	15
Central Visayas							
Mactan	1979	294	294	45		9 695	15
Northern Luzon							
Baguio	1979	156	156	18		12 926	15
Singapore							
All industrial areas[k]	Before 1960	21 994	16 554	6 711		6 804 -35 721	30 [30][l]
Taiwan							
Kaohsiung							
Kaohsiung	1966	168	168	168	10.4[m]	10 333	i
Nantze	1970	222	222	165	37.6[m]	7 257 -10 333[n]	i
Taichung	1971	57	57	57	38.3[m]	4 392 -10 333[n]	i
Thailand							
Bangkok							
Lard Krabang	1980	69	69	46	–	779	i

a) The entire island (160 sq. miles) is a free zone of which three areas are designated as industrial areas: Batu Ampar/Muka Kuning (1 112 acres); Sekupang/Tanjung Uncang/Sagulund (914 acres): Nongsa/East Coast/Bukit Pancur (7 117 acres).
b) By 2004.
c) By 1984.
d) 6 per cent of land value.
e) Only about five acres for EPZ, the rest is for bonded warehouses.
f) Depends on whether buildings are 10 ft. or 20 ft. high.
g) Sites can be made available in three months.
h) 92 for light, 138 for medium and 184 for heavy industries.
i) Land sales frozen.
j) Negotiable.
k) In 1980 there were 19 industrial estates in operation and 15 under development or planned.
l) Also 60 year leases possible in some areas.
m) Payable in monthly installments over ten years.
n) Depends on whether in ready-made buildings or land for own factory.

Note: All figures refere to data around 1980 converted into US dollars with following exchange rates: Indonesia – 627 Rupiah/$; Malaysia – 2.18 Ringits/$; Philippines – 7.51 Pesos/$; Singapore – 2.14 S $/$; Taiwan – 36 NT $/$; Thailand – 20.5 Baht/$. These exchange rated used also for all other data.
Source: Spinager "Objectives and Impact of Economic Activity Zones: Some Evidence from Asia". Weltwirtschaftliches Archiv, No. 1, 1984.

3. There were relatively low labour costs;
4. There were the tariff and tax advantages of the SEZs. At the same time, the Chinese made it amply clear what they expected to gain from the creation of the SEZs, namely, some of the explicit goals for EPZs in other parts of the world:
   - the introduction of foreign capital for development of an industrial base;
   - a means of increasing exports, particularly manufactured goods;
   - a means of absorbing local labour surpluses;
   - the possibility of introducing new technology;
   - a means of stimulating the local economy.

An analysis of the success of the SEZs will involve a detailed look at the expectations and the performance of these zones, on the basis of the above criteria.

The EPZ has often served in the past as a location for relatively non-complex assembly work rather than large industrial production bases with high technology. These EPZ assembly industries in other developing countries have been vertically integrated into larger transnational corporations, thus assuring an up-stream and down-stream control of the means of production and distribution. This differs little from the strategies of some of the national trading companies in the 18th and 19th centuries, also operative in what are now called developing countries[6]. Today, where high technology may be required for production, the process is usually carried out in an industrial country, where the whole range of production and assembly work can be accomplished.

The traditional production sectors of EPZs have been in the areas of electronics, clothing, footwear, leather goods, a variety of toy goods, and plastic items. These sectors do not require high technology equipment, nor do they require particularly skilled labour[7].

Investment in existing EPZs has followed a pattern that is adapted to the production structures and capacity of the given EPZ: most investment has been concentrated in relatively basic production units for labour-intensive manufacturing industries located in the zones. This investment rarely surpasses a total of $1 million, and very often is well below $500 000[8].

Foreign exchange earnings from EPZs in developing countries have been disappointing to the many host governments (Taiwan is a notable exception). The foreign exchange earnings for the host economies is what is earned on the domestic value added (the salaries of the workers and the local services expenditures, when those are run by the host economy, and not as a subsidiary of a foreign firm). As most of the foreign firms in EPZs repatriate their profits, or shift them to other operations outside the host economy, it is not possible to evaluate the net foreign exchange earnings by simply deducting the foreign exchange outlays by firms from foreign exchange profits through exports. The internal capital flow of a firm must first be analysed before any exact estimate of earnings remitted to the host economy can be made (this includes depreciation costs which are not necessarily retained in the host economy). The other source of foreign exchange, tax revenues to the host country from EPZ firms, are also low, as the tax holiday principle is one of the incentives offered by the host governments to foreign firms to invest in the EPZ (these can range from one to five years, at the end of which the firm may transfer the operation elsewhere)[9]. Much of the foreign exchange earnings must go to pay off the international loans that have been contracted by the host government to build infrastructure and administrative units. It has been estimated that the majority of export processing zones, exclusive of the Chinese SEZs, each earned less than $10 000 000[10] by 1983 for the host economies.

The attempts to increase the local value added in EPZs have run into difficulties. Due to the favourable duty free import regulations for the EPZs, most firms, already under an international subcontracting agreement, import semi-manufactured goods from abroad where they can secure quality goods at competitive prices. Where the local economy can produce quality, competitive intermediate goods, local value added can be increased considerably[11].

The volume of semi-manufactured and manufactured exports and the share in national exports can vary greatly. In most cases it is below 5 per cent of the national export value. The EPZ, therefore, does not act as a special stimulus to export, although this can vary according to the sector.

The role of the external economy for an EPZ is an important factor in evaluating its performance. In the case of most EPZs in developing countries, significant external economies do not exist; the inward linkages for EPZ manufactured imports to the domestic economy are weak and the skill up-grading of semi-skilled domestic work has not been greatly successful. Production conditions themselves differ greatly between the EPZ and the domestic economy and the spillover effect is less than most host countries had hoped for originally.

Rates of return on EPZs for host governments have varied considerably. It has been argued recently that the privately sponsored EPZs have had a much greater rate of return than those owned and run by host governments. The World Bank estimated cost for construction of an export processing zone of 100 hectares was $25 to $40 million, indicating orders of magnitude for necessary rates of return on investment for host governments[12].

Another option for host governments is to allocate land space to the highest bidders in rounds of bidding taking place at the end of each lease. In this fashion, the appreciated land value for industrial space is recuperated by the host government in the new leases that are granted to firms. It also means that the first firms have relatively lower land costs. In this fashion, the subsidies offered to firms by lease and land costs would be considerably modified.

EPZ Performance Problems

Direct foreign investment is the principal form of foreign investment in the EPZ. The "new forms" of investment, including turnkey operations, machinery imports, and equity joint ventures have been less used in the traditional EPZ. The EPZs' record as a means of up-grading technology and technological education has not been fully evaluated. However, recent studies document trends that suggest EPZ goals have only partially been fulfilled[13].

The typology of industries within existing zones indicates that assembly work is the principal type of employment in the EPZ. The training required for most of the work force within the EPZs is estimated to be between four and six weeks. It is doubtful whether real skill up-grading can take place under such conditions. The majority of goods produced within the EPZ are already destined for known buyers through the system of subcontracting, assuring the static nature of product upgrading[14].

However, the performance of the more successful EPZs indicates that there is an evolution of skill level and product line when the conditions are favourable. The Shannon, Taiwan and Korean zones are good examples of this. The Shannon zone started as an assembly zone, and evolved in a fashion similar to that described above. It slowly moved to high-technology industries and the workforce composition changed, moving from 13 per cent of the workforce in clerical/managerial positions in 1962, to 19 per cent of the workforce in 1975[15].

The case of Taiwan's EPZs is of special importance because the Chinese authorities closely studied its operations before setting up their own SEZs. In terms of the net foreign

79

exchange earnings, the three Taiwanese zones have accomplished the stated objectives: from 1966 until July 1980, the aggregate value of $2.6 billion in foreign currency had been earned (after local costs and imports were paid) for the host government. Most of the service costs that were paid in foreign currency were remitted to the local economy as well (approximately 75 per cent of the total earnings for the three zones)[16]. The three zones also attracted enough foreign investment in the first five years to pay back the initial investment to the host government; FDI was 13 times the Taiwanese capital investment, and eight times the amount in foreign exchange was earned. It is reported that by 1980, 293 firms had made a net capital investment of more than $300 million (as against the $12 million invested by Taipei between 1965-1970, in 1970 constant prices)[17].

The local employment goal was also judged a success by the Taiwanese. By 1980, more than 81 000 people were employed in the three zones, accounting for 2.9 per cent of the national industrial work force[18].

However, PRC scholars looking at the Taiwan performance tended to discuss the low salaries (which they labelled "exploitation"), the fragility of the EPZs in Taiwan vis-à-vis the international economy and the poorer performance of the zones in the recent world recession.

In fact, backward linkages into the domestic economy have been poor until relatively recently[19]. Nor has technology transfer to the domestic Taiwanese economy been successful. The EPZs there have the lowest capital-to-labour ratio of manufacturing industries. The machinery used is not of high technological conception, and is sometimes even outdated for the assembly process used. As Taiwan is not a member of the International Convention for the Protection of Industrial Properties, foreign firms have been more than reluctant to pass on high technology to Taiwanese associates[20].

Yet, the export oriented strategy that the EPZs represent for Taiwan has brought the country unprecedented growth. It is not clear that such growth can continue, given the heavy reliance upon the international trading system to motor development. The entry of the PRC into the EPZ economy will certainly have an effect upon the Taiwanese zones, as similar production sectors are involved. It has been suggested that the Taiwanese zones may have outlived their usefulness, due to the increased standard of living in Taiwan itself (thus decreasing the competitiveness of the EPZs vis-à-vis other EPZs), and the fact that many firms now want to have better access to the domestic economy.

The social conditions within the zones and impact of the experience upon the domestic economy were important factors for the PRC in evaluating the Taiwanese experience. The majority of the work force in the EPZs were young and female – unlikely to create labour problems nor unionize in large numbers. The labour practices cited in both Western and PRC sources indicate that the social impact of labour exploitation could be a major ideological argument for the PRC in not embarking upon a strategy of labour-intensive, export-oriented development, no matter how large or small the share of exports is in the life of the national economy[21].

In China, there emerged a consensus, however, that the EPZ experience in Taiwan was a success for that economy. PRC scholars expressed the feeling that if Asian developing countries were to become independent of industrial countries, they would have to pass through a phase of technology and skills transfers; they stressed that painful as it might be from an ideological point of view, it must be seen as a long-term strategy for full development of the country. In no case, they warned, should these EPZs be seen as a renewal of foreign concessions or enclaves. They must become the tools of new forms of international economic co-operation, managed by the developing countries themselves, but profitable to the industrial countries as well.

It became clear that new types of export zone units would be necessary for the national economy in the PRC if the policies of opening to the outside world were to be effective. It is in this light that the Chinese authorities turned to a discussion of the creation of the SEZs, based upon their understanding of EPZs elsewhere.

CHINESE ECONOMIC PLANNING AND THE SEZs

Since 1949, Chinese planning authorities have continually devised new categories of economic organisation in the country. The taxonomy of these units is more relevant in Chinese than in other languages, but it is useful to differentiate the several terms at the outset[22].

In the early 1960s Zhou En Lai proposed special areas of the country where export facilities would be set up for light manufactured goods. These were called *"chukuo jiangong qu"* in Chinese, a term later associated with Taiwan's export processing zones. These were not actually put into any degree of operation until after the Cultural Revolution; some 27 bases exist in Guangdong, Jiangsu, Shandong, Zhejiang, Liaoning, Hebei provinces, and some autonomous regions. Specialised export factories (98 of them) were also set up. According to recent information (1984), these bases and factories account for 33 per cent of all Chinese export receipts. At least one of these export bases is designed to service the Hong Kong and Macao economies[23].

The strategy for these bases has been to concentrate export activities within special areas where expertise and quality control could be built up. During the whole period 1949-1980, China's export value was small, and due to the closed nature of the domestic economy, it was necessary to concentrate export industries in different areas of the country where they could take advantage of local resources and know-how. Inside the existing export bases, (called "production bases"), 821 enterprises are owned and run by Chinese foreign trade corporations; 130 co-operative enterprises are jointly managed and run by industrial or agricultural trade units. These bases have proven to be largely responsible for the increase in manufactured goods exports in the period 1980-1984[24].

In the Sixth Five Year Plan document (1981-1985), Premier Zhao Ziyang announced the creation of high growth areas which did not correspond to provincial or administrative jurisdiction[25]. These "economic zones" as they were called (the term in Chinese is *"jingji qu"*) would be created to correct imbalances in the rural-urban exchange of goods and services and provide a motor for the Chinese opening to the outside world. The most important of these zones is to be centred on the city of Shanghai. The State Council decided in December 1984 to extend the Shanghai Economic Zone from the then existing limited zone around the city of Shanghai to the entire land surface of Anhui, Jiangsu, Jaingxi and Zhejiang provinces; this will include 49 cities and 301 counties. The new economic zone, which in fact is an economic co-ordination zone with limited administrative and planning powers, and an unclear relationship to existing provincial and municipal authorities, accounts for 26.6 per cent of the total number of industrial enterprises in China and 26.5 per cent of the total gross industrial and agricultural output of the nation[26]. This zone will resemble more an "export commodities base" with an emphasis upon heavy industry and new technologies. The economic zone will be expected to absorb foreign technology and incorporate it into the renovations planned for the large industrial base in the zone. One of the expected benefits of this will be the transfer of appropriately assimilated technology to the internal economy. Another such zone is projected for the Pearl River Delta area centred around Guangzhou.

Foreign experts have projected nine areas of priority foreign investment for China; these could, it is assumed, become "economic zones" if they are not already designated as such[27].

Related to the separate economic status accorded above to selected areas of China, is the related problem of the status of Taiwan and Hong Kong. The term used for a proposed special status for Taiwan and Hong Kong, for example, is "special administrative zone" *(tebie xingzheng qu)*, a quasi-autonomous administrative and economic status which would allow for the continuation of a modified capitalist economy within the framework of the socialist Chinese state[28]. This remains, of course, highly theoretical and there is no indication that authorities on Taiwan are anxious to subsume this identity within the People's Republic, despite the assurances from Beijing. The future status of Hong Kong is linked to this category of "economic zone" and "special administrative zone", both of which are now being developed in legal terms by Chinese authorities.

This special category of "administrative" and "economic" zones is undoubtedly an interpretation of Article 31 of the 1982 Constitution[29]. Deng Xiaoping has further elaborated the concept in announcing the doctrine of "one country, two systems" when talking about the future of Hong Kong within the Chinese economy. The above categories should be seen as the legal framework for differentiating the socialist state into distinctive planning units with different economic and legal options, all within the single Chinese state.

In addition to the production basis, and the evolving concept of "economic zone", Chinese authorities have created two other areas specially designated for foreign trade and investment. The first in time was the "Special Economic Zone". The second was the concept of a special investment area, limited to coastal cities.

In August 1979, the Central Committee of the Communist Party of the PRC decided on the creation of the first two SEZs in Guangdong Province. In October of the same year, two other zones were created, one in Guangdong and a second in Fujian Province. The SEZs were to be part of the open door policy devised by the team around Deng Xiaoping, aimed at moving China closer to the mainstream of the international economy. The SEZs were to be special instruments of that policy: these new enclaves would be areas in which foreign direct investment would be permitted for the first time in more than thirty years; the zones were to be first and foremost a means of attracting needed technologies for China's modernization. Aside from the turnkey purchases which Chinese authorities also saw as vital to the effort to build up industry, new technologies and equipment would have to be attracted in the form of equity shares of joint ventures, or through co-operative agreements (co-production, compensation trade, processing and assembly agreements). At the same time, new management techniques and production processes could be brought into China – in limited places – without opening the whole of the economy to trade liberalisation and deregulation. They would also be modelled after existing export processing zones, in Asia and elsewhere.

The SEZs were discussed at the provincial level intensively during the period September 1979 to August 1980, when the Fifth National People's Congress approved the Regulations on Special Economic Zones in Guangdong Province[30]. At first, the legal structures which were issued covered only the Guangdong SEZs although it is understood that these are the prototypes of legislation for other such zones in China and that legislation governing the Guangdong zones would apply to all designated SEZs. By January 1985, similar regulations were already published by Fujian Province for Xiamen Zone[31]. It is useful to quote the text of Article 1 of the law as it gives the rationale – albeit in a rather bald form – for the SEZs:

> Certain areas are delineated from the three cities of Shenzhen, Zhuhai, and Shantou in Guangdong Province to form SEZs (*jinji tequ* in Chinese) in order to develop external

economic co-operation and technical exchanges and promote the socialist modernization programme. In the special zones, foreign citizens, overseas Chinese, compatriots in Hong Kong and Macao and their companies and enterprises (hereinafter referred to as investors) are encouraged to open factories or set up enterprises and other establishments with their own investment or undertake joint ventures with Chinese investment, and their assets, due profits and other legitimate rights and interests are legally protected.

Article 4 focuses on the conditions and goals of the SEZs:

In the special zones, investors are offered a wide scope of operation, favourable conditions for such operation are created and stable business sites are guaranteed. All items of industry, agriculture, livestock breeding, fish breeding and poultry farming, tourism, housing and construction, research and manufacture involving high technologies and techniques that have positive significance in international economic co-operation and technical exchanges as well as other trades of common interest to investors and the Chinese side, can be established with foreign investment or in joint venture with Chinese investment.

The map in Annex 5 gives the location of the four SEZs[32]. Shenzhen lies behind Hong Kong; Zhuhai is behind Macao; Shantou is a small coastal city to the north of Hong Kong. Xiamen, which was created separately, was at first limited to Huli industrial district on the island of Xiamen, near the old treaty port city of Amoy (Xiamen); in 1984, the central government extended the SEZ to the whole of the island. Later in 1984, Shantou and Zhuhai zones were also extended to their present size.

The SEZs of China are different from the traditional free trade zones. Originally, the latter functioned as enclaves within domestic economies directed towards export markets. Export processing zones generally had few "backward" links of either goods or technology[33]. While they were free of customs duties, their purpose was to facilitate entrepôt trade. As time went on, the free trade zones also acquired functions for handling customs clearance and relay of merchandise to internal domestic markets, although this was not a principal activity.

Export processing zones were created in many developing countries as a means of sustaining an export-oriented development strategy[34]. These zones all had similar characteristics: exemption from customs duties on imported raw materials and manufactured goods, as well as equipment and machinery needed for export production, fiscal advantages for investors from foreign countries, etc.

The host governments sought specific, if not unrealistic, advantages from the EPZs: foreign exchange earnings, transfers of foreign technology and management skills, new labour skills, some creation of employment locally as well as some spill over of economic benefits to the domestic economy. In the case of the PRC, the programme of modernization required that China have access to advanced techniques, medium to advanced technologies, new management skills and foreign currency reserves to pay for the transfers of know-how and equipment. While this need became apparent during the more euphoric period of Chinese planning in 1978, the realities of the world economic situation and the internal slow-down of the Chinese economy made it clear by early 1979 that "readjustment" would have to be enacted to save the post-Mao economy from an excess of investment. Therefore, the SEZs became a useful tool in both the programme of Modernization and that of "readjustment" as it allowed the central government to restrain the national economic forces and limit more stringently over-investment and over-consumption while at the same time opening the country up to the forces of foreign economic planning in isolated "economic zones". In short, open the country at a given speed in given areas. The SEZs were to be control situations, laboratories for experimenting with the use of capitalist methods and management, while at the same time

83

a means of transferring – in the best of scenarios – capital goods and equipment, in the form of advanced technology into the Chinese economy at large. The SEZs might also serve as a means of testing and reforming the Chinese economic system itself, including attempts at understanding and integrating business cycle phenomena and giving new content to the "law of value" that the Chinese had construed from relevant Marxist literature. The demonstration effect was also a potent incentive to Chinese leaders. The on-the-job economic lessons that were assimilated by the Chinese work force could then be transferred inland to national enterprises and industries. The official literature on the zones stresses the durability of the decision to open up the SEZs to the outside, as well as the important goal of building and maintaining a policy of self-sufficiency in the PRC[35]. Other considerations motivated the creation of these zones as well, including the political solutions for Hong Kong and Taiwan; the zones, argued some, were enclaves of experiment that would show the reluctant comrades from overseas that China was indeed willing to accommodate new styles of economic and administrative life within her borders[36]. This has evolved into the "one country-two systems policy" of Deng Xiaoping.

The initial arguments in favour of the creation of the zones were well publicised. Less well vaunted was the dissension that such a decision caused within the ranks of the Party in the early part of 1980[37]. Powerful political forces were at work to ensure the rationalisation of the economy; and when Gu Mu, a senior member of the State Council, was placed at the head of the special bureau for the administration of the SEZs in Beijing in 1980, such an appointment at once endowed the SEZs with an important political sponsor and a senior economic advisor in the government. The SEZs, in short, were to be one of the major experiments of the post-Mao regime, and it is clear from recent literature that the major justification being given to the different elements with the Party and the government is that to compete in the modern interdependent world China will have a great deal to learn from other nations. By concentrating, but by no means limiting the contact to the SEZs, the government can hope to draw the maximum of advantages from foreigners with the minimum of contact with the Chinese population at large[38]. This was summarised recently in an official article written by the Mayor of Shenzhen:

> There are many advantages to setting up SEZs: Through preferential policies these zones can use large amounts of foreign investment in a better way, import advanced technology and acquire scientific techniques and management skills – all of which will enable the country as a whole to develop economically at a quicker pace. By dealing regularly with foreign capital, we can further observe and understand the development of and changes in the modern capitalist world, and keep abreast of the changes on the international markets and in science and technology. Through co-operation, we can learn modern urban construction management methods, and train professionals[39].

Elsewhere, Gu Mu has made important statements clarifying the innovative – but limited – role of the SEZs[40].

In April 1984, the Chinese government announced a new type of zone to be created for 14 coastal cities that were to be opened to foreign investment. All of the newly opened cities were traditional ports; some were large industrial cities with previous experience in foreign direct investment (Shanghai, Qingdao, Dalian, Fuzhou, Tienjin).

The decision to open the coastal cities was accompanied by statements that incentive packages similar to the investment incentive packages of the SEZs would be created. Instead of a "special economic zone" however, the 14 coastal cities were to have "economic and technical development zones" (*ji shu kaifa qu* in Chinese). These zones are in fact the old industrial areas of the coastal cities that have continued to function as industrial production bases.

Although the incentive packages for the 14 coastal cities are not completely fixed, it is clear from the announced fiscal and land use incentives that the SEZs are still the most attractive cities for investment among those now open to foreign direct investment. This situation could change radically should the municipal governments of the 14 coastal cities, and Tianjin and Shanghai, in particular, be able to modernize their infrastructure quickly. As these cities have the most experience with foreign trade and have basic services already installed, they have potential comparative advantage for massive export efforts in the future.

Basically, the economic and technical development zones (ETDZs) within the 14 coastal cities enjoy the same preferential treatment for most items of the SEZs. However, the labour, land utility and service costs to the firm within the zones are not known. As the ETDZs are industrial areas with functioning enterprises, much of the sought foreign direct investment is aimed at creating import substitution capacity rather than export capacity. This is one of the major differences with the SEZs, which were created to promote export capacity. An important incentive therefore for the foreign investor in the 14 coastal cities is the more generalised access to the domestic market for products; this incentive perhaps more than makes up for the slightly less favourable fiscal and service packages offered to foreign firms within the newly opened cities.

The historical centres of the 14 coastal cities were also set aside as a possible area for investment. However, these districts do not enjoy the same favourable tax treatment as the SEZs or the ETDZs. Basically, the historical districts will have similar treatment to the conditions prevailing for foreign investment throughout the rest of China. The principal differences lie within the tax incentive package. In the historical centres of the open cities, income tax is set at 30 per cent of profits. Enterprises with investments of over $30 million, or who bring in "high technology" as yet to be defined, or who engage in know-how transfers, may apply for a reduction of income tax (to 15 per cent). This in practice will be unlikely, due to the very high investment threshold, and the unrealistic technology transfer demands set by the Chinese initially.

The commercial and industrial consolidated tax is not charged on imported production equipment or goods.

Although the 14 coastal cities have been opened only recently, there are a few joint venture enterprises that were established before the preferential code was promulgated. The experience of at least one of these joint ventures indicates that problems exist in carrying out the incentive package. It was reported that China Schindler Elevator Corporation Ltd., a Shanghai-based domestic market oriented equity joint venture has found it difficult in practice to pay dividends to its foreign partners in hard currency, despite its net profit of Rmb 2.61 million in 1983. This problem of repatriation of profits will be a continuing concern for foreign partners within the open cities. Unlike the SEZs, where foreign currency gains by joint ventures are held at local branches of the Bank of China, open city joint ventures will not only remit a portion of foreign currency earnings to the central government, but they will be obliged to use special accounts in the Bank of China to handle their earnings. This situation seems to be evolving rapidly, with the news that foreign banks may undertake limited commercial operations (and notably joint venture foreign currency deposit accounts) within the coastal cities.

HONG KONG AND THE SEZs

Due to its proximity to the Shenzhen SEZ, and the great importance that Hong Kong has as an external economy for the zone, it is important to understand the recent economic development of the British-held territory to fully comprehend the policies being implemented inside China's SEZs. This is all the more relevant, in that Shenzhen was created partially as a means of demonstrating to the Hong Kong government that China was indeed serious in her attempts to modernize the economy; it was also a signal that the PRC was just as serious in wishing the eventual integration of Hong Kong into the PRC after 1997.

CHINA AND HONG KONG

Hong Kong was ceded to Great Britain in the 19th century to serve as a base for commercial operations with China. The trade activities set up in the last century remained basically unchanged until the end of the 1950s when the territory began to feel the impact of the momentous changes that were taking place on the mainlaind of China. Hong Kong began developing service sectors related to the entrepôt trading functions which previously formed the mainstay of economic life: ship repair yards, refuelling ports, merchant houses, insurance companies and banks[41].

The change of political regime in China at the end of the 1940s caused flows of capital and skilled manpower into the British-held territory especially from cities like Shanghai and Tiansin. The embargo on technological goods trade that the United States imposed upon China as a result of the Korean conflict was strictly applied by the Hong Kong authorities, triggering the beginning of the decline for entrepôt trade with China. These measures forced Hong Kong to define a new economic strategy, based upon the development of light-industrial goods production for export purposes. Within thirty years, Hong Kong built a powerful light industrial base for international export.

By the early 1970s, China was once again looking for means of increasing trade with the region; after formal diplomatic relations were established with the United States in 1979, capital goods once again began to flow through the port of Hong Kong towards China. Merchant banks began to establish themselves in the territory. The authorities in Hong Kong relaxed fiscal regulations on foreign deposits, making it possible for Hong Kong to rival Singapore as a financial centre, especially in the Asian dollar market.

During the period 1970-1984, links between Hong Kong and China increased dramatically. PRC banks (see Table 22) became more involved in Hong Kong's financial affairs. Large trade surpluses were recorded with China; very high levels of remittance towards the PRC flowed through Hong Kong (see Table 23); Chinese-owned banks diversified their activities in the territory to meet new investment priorities.

The PRC state owned Bank of China (BOC) has a large office in Hong Kong, as do twelve "sister banks" of the Bank of China. These 13 banks together form what is called the Chinese State Controlled Banks Group (CSCB). By the end of 1981, CSCB had more than 193 branch offices within Hong Kong, giving them second place as a banking network in the colony (behind the Hong Kong Group).

Besides the CSCB network, the Chinese authorities had control of 13 wholly-owned deposit taking companies, five insurance companies and two joint venture merchant banks (at

Table 22. CHINA STATE CONTROLLED BANKS

Name	Status
Bank of China	
Bank of Communications Kwangtung Provincial Bank	Government controlled banks even during the nationalist era.
China and South Sea Bank China State Bank Kincheng Banking Corporation National Commercial Bank Sin Hua Trust, Savings and Commercial Bank Saving and Commercial Bank Yien Yieh Commercial Bank	Private banks taken over by the communist government in 1949-1952. They are now in effect part of the People's Bank of China. They have lost their identities in mainland China but are allowed to retain their original names in Hong Kong.
Chiyu Banking Corporation Hua Chia Commercial Bank Nan Yiang Commercial Bank Po Sang Bank	Incorporated under the Hong Kong Banking Ordinance with Hong Kong or overseas, Chinese residents holding minority equity shares.

Source: Youngson, *China and Hong Kong: The Economic Nexus*, p. 32.

Table 23. CHINA'S NET FOREIGN EXCHANGE EARNINGS FROM HONG KONG

Unit: $ million

	1977	1978	1979	1980	1981	1982	1983
Trade surplus	1 741	2 259	3 034	4 407	4 116	3 944	3 975
Remittances and other unrequieted transfers	395	477	563	674	n.a.	n.a.	n.a.
Travel and tourism	223	367	819	952	n.a.	n.a.	n.a.
Investment profits	368	461	610	825	n.a.	n.a.	n.a.
Total	2 728	3 565	5 026	6 858	n.a.	n.a.	n.a.

Sources: 1. *Nexus*, p. 58.
        2. *Statistical Yearbook of China*, 1983.
        3. *China's Customs Statistics*, 1984.

the end of 1981). The net worth of these holdings has been estimated at about $1.6 billion[42]. Prior to 1970, China's Hong Kong dollar surplus was principally used by the CSCBs to build up China's reserves and to finance China trade; no real attempt was made to attract Hong Kong dollar deposits. After 1970, however, Chinese financial institutions began to seek out such deposits; they also began to make loans to Hong Kong based firms that were not directly involved in the China trade.

The CSCB now has activities that include the acquisition of real estate for bank buildings and staff quarters, the development of computerized services for the banking sector in Hong Kong; participation in syndicated loans and participation in real estate development projects sponsored by the Hong Kong government. The group enjoys a monopoly over a number of financial operations such as remittances to China, renminbi accounts, and branch banking within the SEZs. This latter condition may change, as the State Council has been studying the proposal to let foreign banks undertake commercial activities within the SEZs.

Chinese presence in Hong Kong is not limited to the financial sector; real estate holdings also represent a significant investment. There are also large investments in the industrial sector, and an extensive use of Hong Kong as an information centre and means of following the international economic situation. Several central ministries have close contacts with Hong Kong based consultant firms or development companies, and the SEZs of Zhuhai and Shenzhen have representative offices there.

Chinese participation in the industrial sector is difficult to quantify, due to the prevailing practice of using Hong Kong residents as middle men to disguise real ownership. One estimate for Chinese net worth in the non-financial establishments was given as $400 million at the end of 1980. Prior to 1976, PRC visible participation was limited to two firms producing cigarettes and monosodium glutamate. By 1983, a machine tool and heavy machine plant, a shipyard, several textile firms and a number of joint ventures producing electronic goods had been added to the list. Recent estimates of profits indicate that PRC investments in Hong Kong are highly profitable: in 1977, it was estimated that profits were $367 million, growing in 1980 to $820 million (of which $230 million came from the financial sector and $595 million came from the non-financial sector)[43].

China's dealings in the real estate sector are more difficult to assess. Prominent among the development corporations affiliated to and owned by PRC is the powerful Everbright Corporation, in charge of developing the Beilin industrial district in Zhuhai SEZ. The latter has made a significant number of investments in real estate. Some estimates placed PRC holdings in the territory in 1980 at HK$ 10 billion[44]. This share has certainly risen due to the massive Chinese effort to sustain the real estate market during the 1982 slump in prices.

Hong Kong is also being used extensively as an information centre for the PRC authorities. Most of the foreign trade corporations have offices in Hong Kong, principally to monitor international markets. The telecommunications network of Hong Kong is also an attraction, allowing firms quick and efficient links to the world and providing them with high international business profile.

Hong Kong serves as a base for the development companies of several industrial districts in the SEZs as well. The China Merchant Steam and Navigation Company manages the Shekou industrial district in Shenzhen; Everbright, as mentioned above, has a role in the development of the Zhuhai SEZ; the Xiamen SEZ United Development Corporation also has close connections to Hong Kong (it is a joint venture among Xiamen SEZ Construction and Development Corporation, the Trust and Consultancy Division of the Bank of China, four Hong Kong based sister banks of the Bank of China, and one Macao based Chinese owned bank).

Hong Kong's involvement with China is equally developed. In the area of finance and banking, three Hong Kong based banks and one Singapore based bank are permitted to have commercial functions within the PRC (the actual commercial functions of these banks is quite limited due to lack of consumer demand): the Hong Kong and Shanghai Banking Corporation, The Chartered Bank, the Bank of East Asia, the Overseas Banking Corporation. Apart from commercial activities, Hong Kong financial institutions have provided a large amount of China's external capital. It has been estimated that by the end of the first quarter of 1980, China's net liabilities to Hong Kong accounted for about 30 per cent of China's foreign debt[45]. This proportion declined in the following years, due to China's retrenchment programme and the high level of world interest rates. Once these rates fell to normal levels, and as access to preferential markets is limited, it is almost certain that China will make use of Hong Kong's financial facilities.

Following the success of the recent issue of Yen bonds on the Tokyo market by the CITIC (1982), by Fujian Investment and Enterprise Corporation (1983), and by the Bank of China (1984), it is probable that the Hong Kong capital market will also be used as a source of external finance[46].

Hong Kong investment in the PRC is difficult to estimate, as figures are irregular and scattered. Strong evidence exists that Hong Kong is the most important investment source for foreign direct investment. Although the Hong Kong investment share – more than $8 billion for the period 1979-1984 – is quite high, the form of that investment is particular. Hong Kong investors prefer to invest in assembly operations, processing, compensation and barter trade rather than equity joint ventures. They also have a preference for contractual joint ventures for large real estate or tourism projects, rather than industrial joint ventures. Equity joint ventures involving PRC and Hong Kong partners therefore make up only 18.5 per cent of the total for the period 1979-84[47]. However, investment patterns in the SEZs are quite different. Hong Kong and Macao businessmen have been reported to be involved in 90 per cent of SEZ investment contracts[48], and their share in total pledged amount of investment should be in the same order of magnitude.

Hong Kong as a Service Base

Hong Kong's relations with China are not limited to trade, investment and technology; there is also very active participation in the service sector. This has been borne out by a the shift in the structure of the Hong Kong economy towards an increasing share of services and correlatively decreasing share of manufacturing industries. This trend should be more obvious for the years 1983-1984; previous years had registered unusual depression of the real estate and the financial markets because of high interest rates; these two activities constitute more than a third of service sector activities. In the budget speech for 1981-1982, the then Financial Secretary of Hong Kong, Sir Philip Haddon Cave, declared "I envisage that the relative size of the contribution of the tertiary services will continue to increase due in no small measure to the demands likely to arise from the expansion of China's international trade, a condition and a consequence of the Four Modernizations Programme". The range of services that Hong Kong can offer to China is nearly unlimited in scope. Financial services and trade services – which constitute a traditional activity in Hong Kong – have already been mentioned. But the next decade should witness a large increase in business services. Many legal consulting firms are already established in Hong Kong; as China's legal system is not yet stabilized, and as the Chinese seem to prefer contractual joint ventures to other forms of investment, the activity of these firms is likely to increase.

Probably of longer-term importance is the use of Hong Kong as an information centre connected to the major financial markets of the world, thus more and more market and feasibility studies for firms based in the SEZs are conducted by Hong Kong based consulting groups. It is also likely that Hong Kong's strong position in merchant transport – Hong Kong has the third largest multi-purpose transport fleet in Asia, behind Japan and South Korea; it is also the most modern one – will give her a competitive edge to maintain her market share of maritime transport implicating China. However, Hong Kong hopes to become a major service base for southern China offshore activities have proved to be too great: Hong Kong will only receive part of this development, as China developed its own service bases in Shekou and Zhanjiang; and Hong Kong will supply only secondary services such as insurance and telecommunications.

It appears as if the economic interactions between Hong Kong and China will be intensely important for the development of the two economies. Hong Kong's comparative advantages seem to lie more in international finance, business and trade services, rather than its manufacturing industries, since the territory does not seem ready yet for the up-grading of its industrial structures towards higher technology fields. The *laissez-faire* tradition which was in great part responsible for the success of Hong Kong is now a hindrance for the transition to a computer society. Unlike Singapore, the government does not have the political and financial power to instigate these changes and, indeed, businessmen will never permit such a transfer, to launch an activity which seems to require state backing and long-term maturation[49]. Hong Kong will thus be more and more in close competition with China for its traditional exports and will face the alternative of relocating its industries in the adjacent SEZs at the expense of job losses for the territory, and/or to turn to activities in the service sector at the risk of higher vulnerability to world-wide economic cycles. Not only are Hong Kong businessmen involved in small- to medium-sized industrial projects in the SEZs, but also in large real estate and tourism contracts, and even in huge infrastructural projects like the development of Futian New Town in Shenzhen (Hopewell Holdings Limited is to invest HK$ 2.0 billion for the development of Zhuhai SEZ; Gladhover Ltd is to invest HK$ 1.03 billion for the construction of the Zhuhai-Canton-Shenzhen highway).

It is not an exaggeration to say that development of the SEZs has been largely financed by Hong Kong capital; this is all the more the case if Hong Kong hard currency loans to such firms as China Merchants Steam and Navigation and Everbright are taken into consideration[50].

However, Hong Kong investment in the SEZs has followed cyclical trends in the international economy. Thus, in 1982, largely in reaction to the sharp fall in industrial rents in Hong Kong and the slowdown of exports in the territory, Hong Kong investment dropped sharply in the SEZs. With the regularisation of the Hong Kong administrative question in September 1984, investment once again picked up.

The underlying reasons for Hong Kong investment in the SEZs and the rest of China has been analysed in different reports and through many published surveys[51]. It is interesting, however, to review two findings in closer detail.

Tables 24 and 25 report the results of the survey conducted in 1982 among 32 Hong Kong industrial firms for a French research institution. The first table shows the ranking of different criteria for selecting host countries. Firms were asked to give a ranking to each of the criteria (between 0 and 10 with 10 as extremely important). In parentheses, we have given a subjective coefficient, indicating whether this particular criterion applies to the SEZs or to the rest of China, and if so, with what ranking. The coefficient can be read as: "0" (not satisfied at all), "1/2" (can be satisfied), and "1" (is satisfied). From this simple system, it can be seen that the SEZs satisfy about 70 per cent of the conditions put forward by Hong Kong investors, whereas China in general satisfied only about 50 per cent. On the basis of this survey, the preference of Hong Kong investors for the SEZs becomes intelligible. As far as equity joint ventures are concerned, whereas the SEZs have attracted (to June 1984) 34 per cent of China's total contracts, and about 22 per cent of the pledged investment in the PRC, they account for 48 per cent of Hong Kong investment contracts, and 56 per cent of pledged investment by Hong Kong firms. Table 25 confirms that cost-related criteria are comparatively more important for Hong Kong firms investing in China than for the whole sample. Correlatively, market related criteria seem to be less important for Hong Kong firms investing in China than for the whole sample. This is one of the principal differences between Hong Kong investors and other investors in China. Whereas American, Japanese, and European investors in China seem to be motivated mainly by market considerations, identified Hong Kong investors engage principally in relocation projects to take advantage of adjacent SEZ lower production costs. This investment strategy accounts for the recent upsurge in Hong Kong domestic

Table 24. MOTIVATION TO HONG KONG INVESTORS

Motivations	Overall	China	Cat
Shortage of labour supply in Hong Kong	5.6	6.3	CR
High labour costs in Hong Kong	5.2	6.0	CR
To facilitate the export of products to another country	5.0	4.0	MR
High land costs and rents in Hong Kong	4.9	5.3	CR
To open up new markets by directly investing in a country	4.3	2.6	MR
Lack of technical and skilled labour force in Hong Kong	3.8	1.2	–
To circumvent tariffs and quotas imposed by developed countries	3.6	1.5	MR
Defending and/or expanding the existing market by investing there	3.3	2.0	MR
To avoid or reduce the pressure of competition from other firms in Hong Kong	2.8	3.5	CR
To further exploit the advantage of the managerial and market skill of Hong Kong parent firm	2.6	4.0	CR
Diversification of products and/or risks	2.5	2.2	MR
High capital costs in Hong Kong	2.0	1.5	CR
To further exploit the advantage of the technical and production know how of the Hong Kong parent firm	1.6	2.6	CR
As a means of managing the financial assets of parent firm	1.4	0.2	–
Lack of higher levels of technology in Hong Kong	1.1	0.2	–
To transfer obsolete technology from Hong Kong	0.9	2.6	CR
Lack of management power in Hong Kong	0.4	0.3	–

CR = China related criteria.
MR = Market related criteria.
Source: Survey conducted by Institut de Recherche et d'Information sur les Multinationales, Paris; except final column, OECD Development Centre estimates.

Table 25. CRITERIA FOR SELECTING HOST COUNTRIES

Criteria	Score	SEZ coef.	China coef.
Political stability	8.2	½	½
Availability of labour	6.9	1	1
Cheap land	6.8	1	1
Cheap labour	6.8	1	1
Good infrastructure	6.4	0	0
Government efficiency	6.3	0	0
Tax concessions	6.2	1	½
Absence of foreign exchange controls and possibility of repatriating profits	5.6	1	½
Possibility and facility for exporting to the developed countries	5.2	½	½
Availability of technical and skilled manpower	5.0	0	0
Geographical location	4.9	1	½
Language and cultural affinity	4.8	1	½
Availability of material components and semi-manufactures	4.5	1	1
Loan availability	4.3	½	0
Size and potential of the host country market	3.8	1	1
Business and family connections	3.6	1	0
Availability of advanced technology	3.4	0	0

Nota bene: Two last columns, subjective indicators compiled by the OECD Development Centre; values of coefficients:
"0" if the criterion is not met by China;
"½" if the criterion is sometimes met;
"1" if the criterion is met.
Source: Survey undertaken for a project sponsored by the Institut de Recherche et de l'Information sur les Multinationales, Paris.

exports to China, as some observers have already noticed[52]. This is probably why Hong Kong businessmen have not been as fussy as their American and Japanese counterparts about the inadequacy of legal business protection in the PRC: they are more interested in mother company commodity flows than in the actual financial performance of the Chinese firm they invest in. This is also why Hong Kong involvement in equity joint ventures has not been very heavy in recent years, as compared to involvement in other legal forms of investment.

Another sample survey of Hong Kong firms investing in China shows that the Chinese counterpart was mainly interested in technology transfer and to a lesser degree in export promotion and net foreign exchange earnings. It appears clearly that the motives of Chinese firms differs from those of Hong Kong firms[53].

Hong Kong and the SEZs

Hong Kong businessmen have frequently complained that labour force problems and the bureaucratic administration in the SEZs have considerably slowed down investment performance. They have, for example, encountered problems with the customs administration, which has been accused of corruption, of making unrealistic demands concerning documentation for imports-exports, as well as taking too long to clear goods through customs. Hong Kong firms have also complained about low labour productivity in the SEZs (variously estimated between 55 per cent, by the OECD Development Centre and 70 per cent by SEZ officials, of that in Hong Kong) due in some part to the number of breaks taken during working hours; on the other hand, the stability of the working force is a strong point for investors, as this is a great problem in Hong Kong.

It is difficult in the present state of available statistics to ascertain the exact contribution of foreign investment to foreign exchange earnings in China and/or the SEZs. It has been reported that net foreign exchange earnings in Shenzhen amounted to HK$ 80 million in 1982[54]; however, these statistics must be treated with some caution, as no indication is given of elements taken into account.

However, it can be safely assumed that the net foreign exchange flow in China due to foreign investment is probably positive if only because profit remittance by the foreign partner is taken out of the foreign exchange account of the venture. More controversial is the impact of foreign investment on technology transfer. Although this would in itself require a very careful case study, scattered evidence suggests that in this respect, the impact of Hong Kong on China is limited.

Hong Kong has never been a major source of technology in any sector, but rather a place where stabilized technology has been adapted and used efficiently. Although this fact would prevent any major technology transfer from Hong Kong to China, it has also more positive aspects:

1. Western technology already in use in Hong Kong can more easily be transferred to a developing country as all the efforts to adapt it have already been carried out;
2. The comparative advantage of Hong Kong investors lies probably in management techniques, i.e. in optimising technology that might not be the most modern, but that can still yield high rates of return in the context of a labour intensive production.

Hong Kong investors often deliver to China machines that are obsolete, Hong Kong firms investing in China admitted that in some cases they want to make use of outdated machinery that was taken out of use elsewhere. This can be seen in Table 24 as well as the survey quoted earlier, in the latter, Hong Kong investors state that in only 5 per cent of these cases did the deals involve "advanced" technology; in 36 per cent they involved "intermediate technology" and in 59 per cent obsolete technology.

Hong Kong investors are also highly concentrated in non-productive sectors or in "low" forms of investment (processing, assembly work, countertrade, etc.), where capital outlay is relatively small. This suggests that disembodied as well as embodied technology transfer must be done on a small scale[55].

Hong Kong, and to a lesser extent Macao, have played a leading role in financing the SEZs of Shenzhen and Zhuhai. Hong Kong's capital, expertise and large international experience have offered partners in the PRC with a world view and market *savoir-faire*. The entrepôt trade experience, and the small export processing experience of the British territory have proved valuable to Shenzhen authorities. The opportunities afforded to China by Hong Kong were multiple, and well used: SEZ administration and firms began setting up business offices in Hong Kong to publicize SEZ investment opportunities, and take advantage of the modern communications installations there to accelerate the investment performance of the zones. The success of Zhuhai, can largely be attributed to the proximity and active support of Hong Kong, already so experienced in playing a discreet intermediary role in China's economic development. No account of the SEZs would be faithful to the realities of that experience without highlighting the energies mobilized in Hong Kong for that experiment. In many ways, the Hong Kong business community began the moves to link the territory more closely to the future of China well before diplomats took the initiative in 1982.

NOTES AND REFERENCES

1. For an excellent overview of the economics of EPZs in an Asian context, see D. Spinanger "Objectives and Impact of Economic Activity Zones: Some Evidence from Asia", in *Weltwirt-schaftliches Archiv,* No. 1, 1984, p. 64 ff. A recent article reviews some of the drawbacks of these zones: J. Woronoff, "Export Processing Zones in Asia", *Oriental Economist,* July 1984, p. 14, ff. For an overview of SEZs, see V.F.S. Sit, "The Special Economic Zones of China – A New Type of Export Processing Zone" in *The Developing Economies,* Tokyo, March 1985.

2. United States direct foreign aid to Taiwan started in 1951 and ended in 1965, when a massive effort was made to set up the Gaoxiong EPZ so that Taiwan could become independent of foreign military and economic aid. There are now three EPZs in Taiwan: Gaoxiong (1966), Nanze (1970), Taizhong (1971). See a review of the Gaoxiong performance in "Taiwan's Export Processing Zones and Foreign Investment", Wu Mei-xan, *Bank of Taiwan Quarterly,* (in Chinese), February 1971, p. 211 ff.

3. See for instance the article in *Zhongguo jingji wenti* (China's Economic Problems), "Industrial Production Export Zones in Asia and the Creation of the SEZs in China", collective authorship (in Chinese). No. 6, 1980. For a review of the performance of the SEZs and the EPZs of Taiwan, see G. Fitting, "Export Processing Zones in Taiwan and the People's Republic of China", *Asian Survey,* vol. XXII, August 1982, pp. 732-744.

4. For a recent overview of the question, see J. Currie, *Investment: The Growing Role of Export Processing Zones, London, (EIU), 1979; Foundation, Free Zones in Developing Countries Expanding Opportunities for the Private Sector.* (USAID, Washington, D.C., November 1983) and the UNCTAD study "Export Processing Free Zones in Developing Countries: Implications for Trade and Industrialisation Policies", Geneva, *TD/B/C/2/211,* January 1983. The USAID study has a selective, up-to-date bibliography. See also, D. Germidis, *et alia, Export Processing Zones,* Paris, OECD, 1984.

5. UNCTAD Secretariat, *TD/B/C.2/211* (Summary), 1983, p. 5.

6. See especially H.B. Morse's, *The Chronicles of the East India Company Trading to China,* 1635-1843, Oxford, 1926. The trading settlements of the Company were in fact prototype free trade zones, and every effort was made to control the total process of production from raw materials to manufacturing through distribution.

7. Total employment in EPZs in developing countries in 1981 was estimated at one million persons. According to UNCTAD studies, the efforts to raise qualifications in these EPZs by on-the-job training have given disappointing results. See the USAID *Free Zones in Developing Countries.* Also UNCTAD, *op. cit.*

8. At the unit level, for instance, capital investments of more than $5 000 per workplace are unusual, with the range being between $1 000-2 000. See UNCTAD, *op. cit.* p. 5.

9. "Where taxes are paid, manipulation of transfer payments can reduce taxable profits of subsidiaries in EPZs". UNCTAD, *op. cit.,* p. 5.

10. "This estimate is based on the average domestic value added to each product in the host country, which is ... below 25 per cent of the value of gross exports", USAID Paper 8, *op. cit.* p. 6.

11. This was notably the case for Taiwan, Malaysia and Korea.

12. Devries and Goderez, *Export Processing Zones,* (World Bank Occasional Paper), SEC M78-612, p. 18. The cost of extending the preconstructed factory space by 100 000 square meters could double the total project costs.

13. O. Kreye, "Export Processing Zones in Developing Countries" *UNIDO Working Papers on Structural Changes,* No. 19, August 1980 (UNIDO/ICIS 176).

14. See USAID *Free Zones in Developing Countries, op. cit.,* p. 9.

15. T. Kelleher, *Handbook on Industrial Free Zones,* UNIDO/ICD.31, 1976, p. 62, as cited in USAID, *op. cit.,* p. 20.

16. Fitting, *op. cit.,* p. 733.

17. Teng Chao, "Brilliant Business Achievements of the Export Processing Zones" in *Jiagong chukou chu jian xun* (Export Processing Zone Concentrates), Taipei, November 1980, (in Chinese).

18. Teng Chao, *op. cit.,* p. 54.

19. By 1980, the domestic Taiwanese economy was the second supply source for the EPZs, after Japan. Cf. Teng Chao, *op. cit.,* p. 12-13.

20. In contrast the PRC, however, passed its first patent law on 16th March 1984, and it went into effect on 1st April 1985. This law was passed precisely to conform to the imperatives of the drive for modernization in ensuring that foreign firms are protected adequately.

21. Cf. Fitting, *op. cit.,* pp. 735-736 for an evaluation of the Taiwanese situation; and *Zhongguo jingji wenti,* No. 34, 1984 for an evaluation of the PRCs view of labour exploitation issues.

22. For an overview of the different types of Chinese economic zones, see B. Louven, "Die Wirtschaftssonderzonen der Volksrepublik: Entwicklung und Modernisierung" in *China Aktuell,* November 1983, pp. 682 ff.

23. Foshan, south west of Guangzhou, supplies vegetables and fish for export, as well as light industrial goods. During Imperial times, the city was a famous porcelain export base.

24. For a comprehensive review of export bases and their place in the overall export strategy, see Tan Qinfeng, "Export Boom Hinges on Production Bases" *Intertrade,* March 1984. p. 35. ff.

25. See "Report on the Sixth Five Year Plan" by Zhao Ziyang, in *Fifth Session of the Fifth National People's Congress,* Beijing, 1983, pp. 113 ff. Zhao states "Apart from Guangdong and Fujian Provinces which can continue their special policies and flexible measures, we should grant more decision-making power to Shanghai, Tianjin and other coastal cities so that they can utilise their favourable conditions to turn their initiative to better account in importing and assimilating technology making use of foreign funds, ... and developing the role of these cities in the world market", p. 131.

26. *Xinhua* News Agency, English press release, 1st July 1983, cited in Louven, *op. cit.,* p. 695. "Shanghai Jingjiqu guoda dao sisheng yishi" in *Jingji ribao* (Economic Daily), 17th December 1984, p. 1.

27. See *China Business Review,* September-October 1983. It now appears clear that no new SEZs will be created, that selected incentives will be extended to the 14 coastal cities that were opened for foreign direct investment.

28. As announced by Ye Jianying; see Yu Guangyuan "Shenzhen SEZ" *Jingji yanjiu* (Economic Research), No. 6, June 1981, pp. 62 ff. 9.

29. Article 31 reads: "The State may establish special administrative regions when necessary. The systems to be instituted in special administrative regions shall be prescribed by law enacted by the National People's Congress in the light of the specific conditions" in *Fifth Session of the Fifth National People's Congress* (December 1982), Beijing, 1983. The future status of Hong Kong, designated a "special administrative zone" after 1997, will be the first use of this new category of legal administration.

30. See "Laws and Regulations Concerning External Economic Relations" for the text of laws adopted between 1979 and 1983 in *Guide to Foreign Economic Relations and Trade,* Hong Kong, 1983, pp. 187 ff.

31. See "Preferential Measures for Investors in Xiamen", in *China Economic News,* 14th January 1985.

32. See *Guide, op. cit.,* p. 209, the City of Xiamen.

33. For an explanation and quantification of these backward linkages in selected Asian export processing zones, see D. Spinanger, "Objectives and Impact of Economic Activity Zones: Some Evidence from Asia", in *Weltwirtshaftliches Archiv,* No. 1, 1984.

34. See, for example, the discussion of the problem in the UNCTAD document *TD/B/C.2/211 1983* "Export Processing Free Zones in Developing Countries: Implications for Trade and Industrialisation Policies", UNCTAD Secretariat, Geneva, 1983.

35. See a typical presentation in the official brochure produced by the Shenzhen municipal government, 1983, p. 3: "The establishment of SEZs is a long term principle for the People's Republic of China to adopt open-door policy and develop external economic and technical exchanges, with the aim of promoting economic construction".

36. See for example the article (in Chinese) of Sun Ru: "The Role and the Construction of the SEZs from a Strategic Point of View" *Xueshu yanjiu* (Academic Research), No. 4, 1982. See also an earlier seminar paper translated into English by the Joint Publications Research Service (JPRS), No. 75423, 2nd April 1980, p. 46.

37. See accounts and echoes of this in 1982 in: "First Approaches to the Problems involved in the Creation of the SEZs in China" (in Chinese) by Zhang Hanqing, *Xueshu yanjiu,* No. 4, 1982 and in the front page article of the national daily, *Renmin ribao,* 13th September 1982 (in Chinese).

38. Liang Xiang, "China's Special Economic Zones", in *Beijing Review,* No. 4, 23rd January 1984, p. 24 ff. Xiang remarks "To offset their weaknesses, developed and developing countries are bound to learn from each other's strong points through extensive economic co-operation and technical exchanges. One form of international co-operation is the special economic zone ... As a developing socialist country China relies mainly on its own forces to bring about modernization. Of course, it will try to win foreign assistance, actively develop economic co-operation and reasonably use and absorb all foreign things useful to it. For these reasons, China has gone ahead with its decisive policy to set up SEZs". Mr Xiang is the Mayor of Shenzhen and the Vice-Governor of the Province of Guangdong.

39. *Op. cit.,* p. 25 f.

40. "Setting up the SEZs and opening the 14 coastal cities are an important part of the open-door policy. Their coming into existence had caused doubts and concern in China and abroad because

SEZs had not been heard of in the Marxist classics. Many asked what they were all about, or whether they would create infringement on China's sovereignty. The fact is: SEZs have nothing to do with infringement on sovereignty. They are not special in a political sense like the proposed formula for Hong Kong and Taiwan which come under the "one country, two systems" concept.

The special economic policies and systems of management in the SEZs take effect under the leadership of the Chinese Communist Party. The task of building up a socialist spiritual civilisation also applied in the SEZs. It should be said that the establishment of SEZs is an extension of Lenin's principle of enlisting capitalist tools to build socialism.

The development of the SEZs will rely chiefly on foreign financial resources. With the limits defined by the Chinese government, various forms of economic entities are permitted to exist, such as joint ventures, joint management, wholly foreign owned enterprises, etc. Their products are mainly for export. Although SEZs are still guided by the planned economy, the market force is allowed to play a predominating role. In other parts of China, it is the planned economy that dominates the economy.

In other respects, SEZs differ from interior regions in giving foreign investors special treatment in taxation, fees for use of land, simpler entry-exit procedures, etc.

The greater autonomy granted by the state to the SEZs is for no other purpose than to provide the kind of investment environment that would encourage foreign investors. However, foreign businessmen must abide by Chinese laws and regulations. This is a fundamental difference from foreign concessions in the old China.

The question has also been raised why it is necessary to set up SEZs when China was already implementing the open-door policy. The answer is that the open-door policy entails a host of practical and palpable methods. The four SEZs were selected by virtue of their advantage in geographical location. Shenzhen's proximity to Hong Kong gives it an edge in bringing in new technology and access to international economic information.

The SEZs serve as a testing ground of China's economic reforms. Some decisions in the current reform have extremely wide implications, and must be treated with the utmost caution. Testing them first in the SEZs before applying them elsewhere can ensure that only the successful ones are introduced in extensive areas. This is not only condusive to searching for new ways, but also has the advantage of keeping the effect of wrong decisions within bounds. In fact, the SEZs act as filters in the course of absorbing foreign management methods.

A case in point was Shenzhen's successful experience in designing and opening construction projects (including the practice of opening construction projects for tenders rather than arbitrary assigning of jobs) and this was subsequently introduced across the country.

The 14 coastal cities, the four SEZs, plus Hainan Island are the forward positions established to bring the technology, information, capital and talents needed by the vast heartland of China in its modernization". Gu Mu, "Why SEZs and the 14 Cities", *Ta Kung Pao*, 29th November 1984.

41. A. J. Youngson (Ed.) *China and Hong Kong: The Economic Nexus*, Hong Kong, 1983.

42. Youngson, *op. cit.*

43. Youngson, p. 49. Other estimates made by Hong Kong Trade Development Council Research Dept. in *Hong Kong Economic Relations with China*, Hong Kong, August 1983 evaluates total Chinese investment in the range of $3-5 billion.

44. Youngson, *op. cit.*, p. 47.

45. Youngson, p. 30.

46. In January 1985, it was reported that the CITIC planned to issue HK$ 200 million certificates in Hong Kong during the first six months of 1985. *Ta Kung Pao*, 31st January 1985.

47. Development Centre OECD database on equity joint ventures in the PRC. The Hong Kong participation is about $130 million.

48. *Economic Intelligence Unit,* No. 2, 1983 (China, North Korea).

49. A sales tax was proposed to Hong Kong businessmen by the government to finance research and laboratories in the field of computer sciences. This proposal was flatly rejected.

50. Everbright's development projects for Zhubai are financed largely through loans to the company by Chinese-owned Hong Kong banks.

51. As reported in *China Reconstructs,* September 1984, p. 17.

52. Youngson, *op. cit.*

53. Such as the surveys carried out by the Hong Kong Chamber of Commerce among its member firms (1980); a survey undertaken for a project sponsored by the Institut de Recherche et d'Information sur les Multinationales (1982); and a survey undertaken by Chai reproduced in *Hong Kong Nexus,* 1983.

54. "Guangdong External Economic Activities" in *SWB/FE/W1223/A/12,* 16th February 1983.

55. Cf. Youngson, *op. cit.,* p. 132 ff.

Chapter 4

THE SPECIAL ECONOMIC ZONES

PLANNING THE SPECIAL ECONOMIC ZONES

From the beginning, the SEZs were planned to have been under separate administrative and political tutelage. In the central government, a special office was created attached to the state council, headed by the senior economist Gu Mu; this office is responsible for maintaining constant contact with the authorities in Beijing, and intervening when problems arise from friction with other central ministries or state supply companies. This extra-ministerial status for the SEZs has guaranteed them a good deal of political independence and speeded up the process of attracting foreign capital to the zones. At the provincial level, a co-ordinating office (the Provincial Administrative Commission for the SEZs) exists for Guangdong and Fujian provinces. This office relays information to Beijing and helps determine planning co-ordination among the different state and collective enterprises in the province and those within the SEZs. In the case of Shenzhen, the vice-governor of the Province became the mayor of the city in October 1981, as a token of the importance the provincial government afforded the SEZ.

The key organisational unit in each of the SEZs is the SEZ Development Company. This firm is a planning and development enterprise, as well as a negotiations bureau for the foreign and Chinese investors coming to the SEZ. In the case of Shenzhen, the Development Company has the responsibility of real estate development and infrastructure development. The firm is also responsible for locating appropriate Chinese partners for joint ventures, and seeking Chinese capital for investment. The Development Company works in tandem with the municipality, which has the responsibility for the legal aspects of development, and oversees the general planning of the zone, in so much as the planning division of the municipal government uses its new statistical and planning departments for forecasting. The plan for the SEZs is not integrated into the national planning exercise, as the SEZs were created initially to export most of the zone production. As the domestic Chinese market becomes an increasingly important element in the SEZ export share, it is evident that national planning bodies will have to take into account in a more systematic fashion the volume of exports to the domestic market. This planning gap has other inconvenient side effects. In the national economy allocation of resources is made by planning authorities according to priorities within the Plan. This allocation implies an output expectation. The problems of centrally planning an economy the size of China are well documented in recent studies, and it is unnecessary to rehearse them here. However, it should be noted that the SEZs are not organised in such a fashion as to have planned inputs and planned production outputs with government procurement quotas. This means that special priority access to raw materials, intermediate inputs and foreign exchange have to be established for the SEZs in order to function with the fluctuations caused by the zones' interaction with the international economy. The SEZ bureau in Beijing was established to ensure that type of priority, but it is clear that this has caused friction among ministries and foreign trade corporations who have seen their own demands treated with the usual bureaucratic procedures. The SEZ links to the national economy, therefore, have to be seen in the light of scarce resources being allocated first to the enterprises and administrations of the SEZs, and then later to the other high priority areas, such as the new 14 coastal cities.

The SEZ Structure

The local administrative structure of the SEZ follows a plan which was developed for all the zones (Figures 6 and 7).

Figure 6. SEZ ADMINISTRATION
Shenzhen Municipal People's Government

Mayor
3 Vice Mayors
1 Secretary-General

Municipal Government Offices	Police, Courts	Shenzhen SEZ Development Co.	Shenzhen SEZ Construction Co.
		Admin. Office	Planning Department[1]
		Intro. Dept.	Planning Division
		Trade/Service Dept.	Building Control
		Land Dev. Dept.	Survey Department
		Housing, Land Property Co.	Admin. Office
			Architecture Dept.[1]
			Landscape Company
			Housing Company
			Water Supply
			Road Building
			Dev. Consult Co.
			Environ. Protect.
			Import./Export Mtls
			Building Mat. Co.
			Quality Control

1. Departments responsible for planning and land development control.
Source :   China Building Development, 1982/83.

The exact boundaries of jurisdiction for the SEZs have evolved over time. Shenzhen is the only zone that has not had a modification of the original terms of reference for the SEZ circumference. Xiamen SEZ's borders are still not completely fixed. Xiamen allows SEZ-type incentives to investors who invest in Fujian Province outside the zone itself[1].

Figure 7. THE ADMINISTRATIVE HIERARCHY OF SEZs

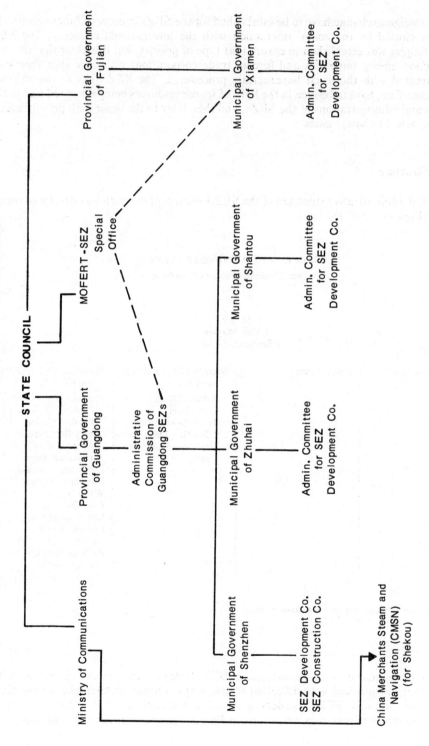

Source : *Guide to Investment in China,* Economic Information and Agency, Hong Kong, May 1982.

Procedures for Negotiations

It would be useful to give a concrete idea of the procedures for negotiating a contract in the Shenzhen SEZ.

Potential investors make arrangements to visit Shenzhen from Hong Kong, as a front office company exists there; the Shum Yip Trading Company is staffed by SEZ authorities. Through this intermediary, investors contact the Shenzhen Special Economic Zone Development Company or the Introductory and Discussion Department, the Special Economic Zone Liaison Company, Guangdong Enterprise Ltd. Foreign businessmen then apply for, and fill in, the Application Form for Investment, and Initiating of Enterprise Undertakings. At the same time, the foreign interests must submit an investment plan with details of intention or purpose of investment, items of investment, scale of investment, land area required, amount of investment, (such as sole investment, joint investment and co-operation) materials concerning feasibility studies. For businesses that are investing in Shekou Industrial Zone, contact is made with China Merchants Steam Navigation; for those wishing to invest in Shahe District, contact is made with the Shahe Overseas Chinese Enterprises Company.

Although the SEZs have created what the Chinese consider streamlined procedures, it is still a rather cumbersome process. There is, however, a better defined line of command in the SEZs than in China at large, as well as a minimum of conflicting interests among official government agencies; the latter has often been cited by foreign businessmen as a major obstacle to negotiation with Chinese officials – one is never entirely sure that one is dealing with the person who has the power to conclude an agreement or to assure delivery of necessary energy or raw materials supplies once the joint venture is underway. Normally, in the SEZs this problem is mitigated by the integrated government and planning structures that form the base of a horizontal co-ordination within the zone. Problems arise, however, when these SEZ specific agencies deal with provincial or central authorities, as is the case for the customs authority (central) or provincial suppliers of energy and raw materials.

Once the forms are submitted with the plans, Chinese authorities have promised to have them processed "rapidly". In practice, businessmen are invited for interviews with the SEZ authority soon thereafter.

The negotiation phase lasts for a number of weeks, or even months. A contract is drawn up by joint negotiation. Once both parties have agreed to the terms of contract, it is passed to the Shenzhen Municipal People's Government for approval.

After the approval of the contract, the two parties work out the detailed list of goods required to be imported, and this is submitted to the Municipal People's Government for examination and approval. Once this procedure has been followed, arrangements are made at the Kowloon customs station for the import licences.

Before actually opening a business in Shenzhen, foreign interests apply to the Industry and Commerce Administration Bureau of Shenzhen for registration, and in this way obtain the business licence. All pertinent documents, including lists of the regulations of the enterprise, the directors of the Board, the copy of the registration certificate and other credentials issued by competent authorities of the country of origin.

The Trade and Service Department of the SEZ takes care of obtaining the business licence and export certificates.

The typical steps in negotiating the *land use phase* of a joint venture might look something like the following:

1. Discuss the project proposal, the choice of site, and the land requirements with SEZ Development Company representatives;

2. Conclude and sign an agreement/contract with Development Company after approval by the Municipal People's Government;
3. Apply for "Land Use Certificate" from Planning Department of SEZ;
4. Approval by Planning Department and issue of "Land Use Certificate" after payment of land use fees;
5. Approval of architectural plans from Architecture Department;
6. Approval of building plan by Building Control Division of Planning Department and issue of Building Permit;
7. Apply for "Construction Permit" and arrangement of construction team from Construction Company;
8. Site preparation and construction.

The above short resumé underscores one of the great advantages for foreign business interests in dealing with SEZ authorities: the red tape and bureaucratic procedures are reduced substantially, when compared to other areas of China. The lines of command and authority are clearly delineated, and the SEZ authorities themselves have direct access to provincial and central government officials without passing through the industrial and trade ministries for approval of contracts. There are also special arrangements for the Bank of China offices in the SEZs, which give SEZ authorities a greater leeway in arranging for Chinese loans or financing for joint ventures within the SEZ.

The Customs authority is the single most important bottleneck in the procedure. Although the SEZ authorities have a clear right to enforce the import-export conditions for raw materials and processed or manufactured goods emanating from production units in the SEZ, in practice there have been numerous problems at the level of the local customs authority, who see their role of watchdog of the frontiers being eroded by special concessions to foreign investors. Efforts are being made to create a smoother operation at the customs level, and once the special frontier has been finished along the northern end of the SEZ of Shenzhen, the problem will be somewhat diminished, although not completely relieved.

SPECIAL ECONOMIC ZONES: AN OVERVIEW OF DEVELOPMENT

The legislation that created the SEZs in China did not set out the exact physical boundaries of these zones, nor did it in fact specifically limit the experiment to the four operative zones of today. The three zones of Guangdong Province (Shenzhen, Shantou, Zhuhai) and the single zone in Fujian Province (Xiamen) were in fact to be forerunners of other economic experiments, should they succeed. In 1979, the State Council also extended a degree of autonomy in economic planning to the island of Hainan in the South China Sea, although it is not officially under the SEZ legislation. Zhuhai and Xiamen have recently begun industrial operations, and they are treated first. Shantou, which is also a medium-sized city has a number of industries already producing. Shenzhen, the largest, and most successful SEZ, will be treated in detail.

Zhuhai SEZ

The Zhuhai SEZ comprises a total area of 15.16 sq. km. located in the lower estuary of the Pearl River, adjacent to the city of Macao, and near the small town of Zhuhai. The initial

102

Figure 8.   INVESTMENT PROCESS IN SEZs

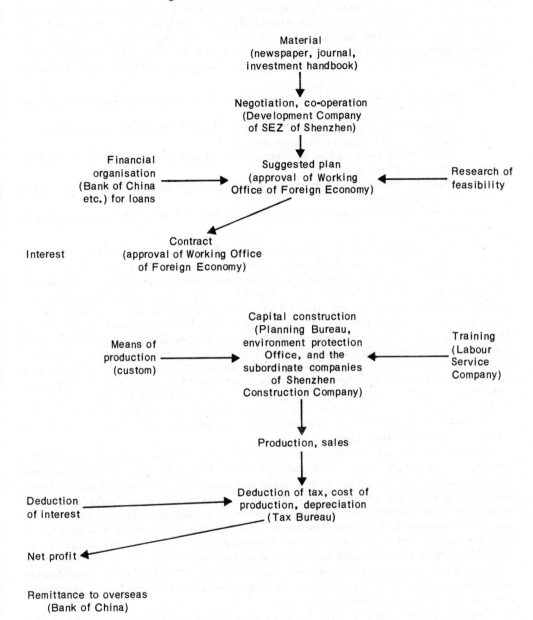

Source : CHU K.Y., "The Recent Reform of Administrative Structure and the Investment Environment of the Shenzhen SEZ" *in* CHU K.Y. ed., *The Largest Special Economic Zone of China-Shenzhen*, Hong Kong, Hua Feng,1983, p. 8.

area of the SEZ was to be 6.8 sq. km., but in 1983 the State Council authorised the extension of the zone to the present limits. The geography of the zone is similar to that of southern coastal China in general, that is to say, alluvial flat areas with agricultural potential. The climate is subtropical and oceanic, with temperatures varying between 3 degrees and 38 degrees centigrade, averaging 22.5 during the year. The humidity factor is high (79 per cent). The initial plan for development called for dividing up the SEZ in the following ways:

a) industrial installations to be located in the central area, and occupying 37 per cent of the land surface;

b) commercial and residential units in the centre and towards the east of the zone, occupying 17 per cent of the land surface;

c) tourist installations (hotels, resorts) in the east and west of the zone, with a land surface of 19 per cent;

d) education, medical and public service facilities are planned to occupy 6 per cent of the land surface;

e) 21 per cent of the land surface is reserved for road infrastructure and reforesting.

The initial plan also called for the completion of infrastructure by 1986, as well as potential extension of the zone into the area of the city of Zhuhai. This was overambitious as events showed. By mid-1982, a number of important service installations had been completed; work began on the Jiuzhou-Hong Kong rail link. Electricity, water and telecommunications services were also installed. A port area of 2.63 sq. km. was cleared and established with the capacity to receive ships of up to 5 000 tons. Thirty-two hotels and residential areas have also been constructed for permanent residents and visitors to the zone. Zhuhai shares with Xiamen and Shenzhen the immediate proximity to a larger external economy, and intends to make use of the Macao area for trans-shipping and tourist facilities. In the final plan, Zhuhai will have more developed port facilities than are presently available at Macao, which has no deep water port. Macao is equally dependent upon the external economy of Hong Kong, and it seems that the future development of Zhuhai will be to bolster and improve the Macao economy, as a complement to it, rather than a replacement for it, thus entering into a complex integrated economy of the Pearl River Delta[2].

The majority of export earnings from the zone come from agricultural goods, including vegetables, flowers, farm animals, fish and fruit. Most of the 1 350 projects that have been signed are small-scale light industry factories or real estate investment for the Macao-Hong Kong community. One example of development in the Yinkeng district of Zhuhai is a project undertaken by the Yianwan Joint Development to build a residential and tourist centre with a capital investment of $300 million. The long-term use of this facility will be as a base for oil exploration in the South China Sea; eventually, a bridge will join the Yinkeng area to Macao itself. The current time frame for the project is seven years (1990).

In order to carry out the necessary prospecting and management of large foreign investment, a zone-wide development company was created with Hong Kong, Macao and PRC participants on the board of directors.

The development plan for the Zhuhai area called for Chinese investment of over Rmb 672.5 million before the year 2000. Chinese authorities expected to attract $227 million in foreign investment during the same period. An early estimation of development foresaw 227 factories in operation with an estimated export capacity of some $568 million by the year 2000[3]. The realities for the SEZ have been quite different. By the end of 1984, Zhuhai had spent more than Rmb 500 million on basic construction; almost the total amount planned for

Table 26. ZHUHAI BASIC DATA, 1984

Location:	Zhuhai district behind the city of Macao in the Pearl River Basin.
Area:	Zhuhai city is 654 sq. km., SEZ is 15.16 sq. km.
Population:	32 800, SEZ: 9 600 (1982 figures).
The economy:	At present, light industrial installations are being constructed. Six major types of export industries are now being built up: building materials, electronics, textiles, everyday chemical products, food processing, engineering equipment and machinery and petrochemicals. Aluminium and porcelain manufacturing are being given special priority. Special emphasis will be placed on tourist resorts.
Foreign trade:	Zhuhai city exports some agricultural products, fish and canned goods. Export volme is not available.
FDI:	Investment since March, 1979 totals $1.5 billion (or 1 200 contracts). Foreign investment in the first quarter of 1984 was reported to be $198 million; the figure for the whole of 1983 was $1.1 million; 80 per cent of the investment was intended for industrial use. Agreed projects include: a glassware factory, a lumber processing plant, the manufacture of microwave telecommunications equipment as well as a refrigerator production line and a brewery.

Source: Intertrade, November 1984.

construction to the year 2000 has already been spent, indicating that authorities have revised their plans several times since the inauguration of the zone. The same holds true for the investment figures. Although the SEZ authorities had initially planned to attract some $227 million by the year 2000, the SEZ had, by the end of 1984, attracted more than $1.5 billion in pledged foreign investment. This corresponds, in part, to the very active role taken by Everbright and Gladhover, as well as the new role of oil service base that Zhuhai may be called upon to play.

Zhuhai would like to attract, in priority, foreign investment in the areas of light industry, agricultural export, breeding, real estate and tourism. It appears that although the central government authority would prefer investment in the areas of industry and tourism, the foreign investors have greatly preferred "non-productive" real estate investment[4].

The most recent sectoral breakdown of figures for foreign investment in Zhuhai bear out the general notion that first phase investment in Zhuhai – like Shenzhen – has been in property development.

However, partial figures for 1983 indicate that 50 per cent of the new investment for the first quarter of that year, was in industrial investment. This trend away from property development into industrial development is characteristic of the other SEZs' second phase as well, where large infrastructure projects and light industrial projects are the favoured forms of investment.

More recent global investment figures (December 1984) for Zhuhai register more than $1.5 billion pledged[5], mostly to be funnelled through a Chinese joint venture, the Jiahao Company of Hong Kong. For the large industrial and tourist projects, the Chinese have made agreements to lease land to foreign concerns. Overall performance for the zone has received somewhat contradictory coverage, but by all reports, foreign direct investment is accelerating[6].

The Chinese wish to acquire new technologies in their joint ventures with foreigners including technology for the production of building materials, especially glass technology for both artisan and construction purposes. However, results to date have been disappointing to the Chinese, as no significant agreements have been reached to supply joint venture facilities with up-to-date technology, due in part to lack of legal protection for patents, the lack of qualified labour inside the SEZ, and reluctance of foreign investors to move directly into large production units.

The infrastructural problem is the key to future development in Zhuhai. A bridge link is planned for Zhuhai-Macao; 14 new roads were built, extending existing road service. Estimated Chinese contributions for this new infrastructure are about $134 million.

Linked to this problem of the development of Zhuhai is the future economic development of Macao. The population of the city in 1983 was approximately 380 000, 98 per cent of Chinese origin. Macao was first settled in the 16th century, when it was used as an offshore foreign base for trade with Canton. Today, the major industrial output there is light industrial goods, principally garment products. Trade figures for Macao show that the China connection is growing stronger; raw materials are being imported from the PRC and Hong Kong for processing and re-export in manufactured form (once again, principally garments) towards industrial markets. This makes the co-operation between the SEZ and Macao all the more important; the Zhuhai zone can serve as an entrepôt and trans-shipping facility for the Macao economy, should it become a major port as planned. To facilitate this shift of export facilities to the SEZ, Zhuhai plans to build a large deep water port in the near future[7].

Table 27.   ZHUHAI: FOREIGN INVESTMENT, JANUARY 1982

Sector	Contracts	Amount pledged ($ million)	Percentage
Industry	191	46.82	19.4
Agriculture	13	4.36	1.8
Commercial	39	8.00	3.3
Real estate	15	164.57	68.1
Tourism	8	15.80	6.5
Other	22	2.23	0.9
Total	288	241.78	100

Source: SEZ Statistical Yearbook, 1983.

The total revenue of services and trade in the SEZ amounts already to more than $100 million since 1982, aside from the expected earnings from the industrial sector of the Zhuhai SEZ. The tourist attractions and capacities of the SEZ are now receiving closer attention. Although the tourist economy was not given a predominant place in the development plan of the zone, it is now evident that the leisure service sector will emerge as one of the big earning capacities for Zhuhai. With the overcrowded urban conditions of Hong Kong and Macao, the new open spaces for recreation promise to be a strong attraction to holidaymakers from the two cities who find leisure conditions cramped and expensive in their own cities; Zhuhai offers bargain vacation facilities for modest income families who cannot afford the package tours that bring so many people out of the territory. The appeal of the zone lies in the unusual countryside (made more unusual by some Chinese developers who have carved mythical figures in the granite rock slab) as well as its "wild game restaurants" (the "yewei"). The tourist complex at Shijingshan has been completed as well as three new tourist hotels. Some of the space in the latter will be taken by oil executives. Tourists to Zhuhai in 1982 numbered more than 360 000, mostly Chinese from Hong Kong and Macao. Hong Kong investors have proposed a HK$ 30 million car race track as well. This type of investment is attractive to foreign partners, as revenues from hotels and tourist facilities are paid in foreign exchange certificates or Hong Kong dollars, both transferable outside the zone with lengthy administrative procedures.

The role of overseas Chinese investment is important. The major foreign investor in Zhuhai is the Hong Kong based Gladhover Ltd., a firm in fact owned by southeast Asian Chinese. British firms are acting as consultants to Gladhover which has also been granted the contract for the new $170 million oil base being built in Zhuhai. A port will be the first phase of the $4 billion investment in Zhuhai (phase I will involve a $70 million investment from Gladhover). When finished, phase I will include the construction of the main harbour and 600 metres of quays[8]. Gladhover has signed agreements for the harbour and the construction of residential estates and two industrial parks. The residential area will be built close to the future site of the Macao bridge, and it will be aimed at foreign oil executives who can afford to rent the high-cost dwellings. As the Zhujiang basin (Pearl Basin) is expected to have the highest commercial oil potential, the two SEZs (Shenzhen and Zhuhai) located on the delta are vying for service base facilities for oil companies. A second contract will involve an industrial park of 170 sq m. in the eastern section of the zone, to which Gladhover will contribute $30 million. Another industrial park will focus on production of clay tiles, glass fibre and building materials; it will be located in central Zhuhai.

In April 1983, the Zhuhai Development Company, the firm that manages the planning of the zone, announced that it was to enter into a joint venture with Macao to reclaim 110 hectares of land in Macao waters. The land will be used for commercial purposes, as well as residential areas, once it is reclaimed. The time-frame on this project is twenty-five years. A second reclamation project is underway in the Areia Preta section of Macao. This will be used for industrial purposes, and will be financed with joint venture capital from Zhuhai and the PRC. It is one of the novelties of the Macao-Zhuhai connection that investment capital flows both ways. Both of these reclamation projects could double the space of Macao in twenty-five years time, and add substantially to the export capacity of the city.

By August 1984, Gladhover had raised $64 million for the Jiuzhou port project, on behalf of the joint venture. The money was secured through a syndicated loan. Ten banks provided the loan. Lloyds Bank International, the Bank of China and the Mellon Bank of Pittsburgh provided $37.2 million for port facilities; $26.8 million was provided for office and residential buildings for the oil base. Preliminary work began in September 1984[9].

Zhuhai hopes to make the best use of its location, both for trans-shipping through Macao, and for servicing oil bases in the South China sea. It is closer than Shenzhen's Chiwan base to oil exploration units in the South China Sea; it is also closer to the open sea for general trading purposes with southeast Asian countries. The port facility, therefore, is a necessary infrastructural investment before any large-scale industrial investment destined for exports can be realised.

In June 1984, the Communist Committee of Zhuhai Municipality adopted revised guidelines to the SEZs investment policies. These new guidelines included measures that allowed joint ventures, under certain circumstances, to sell the products of the SEZ enterprise directly on the domestic Chinese market.

Industries established within the SEZ will be allowed to market up to 30 per cent of their production on the domestic market, if permitted to do so in their initial contract. Gladhover may eventually become more involved in bridging the gap between SEZ production and PRC marketing, and therefore its role as a partner in joint ventures with foreigners could be crucial. The SEZ administration, as well as Gladhover, are looking for an expanded trade role for the SEZ in the region.

Another Beijing owned firm, Everbright Holding Company Ltd., of Hong Kong has been involved in the development of Zhuhai, much in the same way that the China Merchant Steam and Navigation is involved at Shekou in Shenzhen. Everbright entered into a fifty-year joint venture with the Pearl River Water Resources Commission, and the SEZ for the Ma Dao Men Development Company. This latter firm will undertake an agricultural reclamation project of 170 sq. km. on the west of the Pearl River estuary. According to a feasibility study carried out in 1984, the reclaimed land will require an investment of $100 million and six years to complete; the project could then support a large-scale sugar cane production, a sugar refinery of up to 4 000 tons of cane per day, and a variety of vegetable and fruit production that could service the entire Zhujiang Delta[10].

Everbright is also responsible for the development of the Zhuhai Beiling Industrial Estate (4 sq. km.). Twenty factories will be developed in a first phase for lighter industrial use. There are signs that Zhuhai will be able to attract investment in integrated circuit technology and injection moulding[11].

The Zhuhai SEZ is divided, for planning reasons, into six different types of areas, each with a special development focus. The Gongbei district, adjacent to the border with Macao, will be the seat of the SEZ administration, and have the nexus of commercial and trading units centred there; the Xiawan district, to the northwest of Gongbei, will be used for industrial development, and more especially for the building materials industries to be installed. Beiling, a smaller district to the northeast, will be used as the scientific centre of the SEZ, with cultural and research facilities; there will also be light industrial installations for electronics, and textiles. The Jida district has already been mentioned in connection with the Shijingshan tourist resort; the Shihuanshan district, which includes the Jiuzhou harbour and the heliport, is the transport hub of the SEZ. The Silver Valley development is opposite the city of Macao and will be used as a tourist and recreation area, as well as a residential area.

The SEZ is governed by an Administrative Committee, which works under the supervision of the Zhuhai Municipal People's Government. Like other SEZs, Zuhai has a development company that co-ordinates investment. It is also the company through which the municipality makes investments in the SEZ. The Zhuhai Special Economic Zone Development Company already owns or participates in ownership of over 30 enterprises in the SEZ.

Gladhover has entered into agreements with the administration of the SEZ to develop approximately half of the SEZ. This ambitious plan for development is called the Silverbay Development Project. Among the projects set for construction within this plan are the Jiuzhou Harbour, which will be the commercial port for the SEZ as well as an oil base.

Everbright and Gladhover are both developing industrial parks within the Zhuhai SEZ, but with their different emphases they have a minimum of competition for foreign investment. The Gladhover harbour project and the Everbright agricultural project are largely complementary as well, and will bring economic benefits to the whole SEZ. The chief form of competition for FDI will come from Shenzhen and several of the newly opened 14 coastal cities.

108

Xiamen

Xiamen, unlike the three other SEZs, is located in the Province of Fujian, along the China coast. The city has a population of 969 000 (1984), and a total land area of 1 515 sq. km., making it the second largest city in the province. The traditional economy of the city was centred on fishing and trade with the inland provinces of China, and during the 19th century the port became an active export-import centre for the rest of China. Xiamen is better known to the Western reader under the name of Amoy[12]. Xiamen was one of the principal ports under the Ming and Qing dynasties, and today it maintains trade relations with more than 60 foreign countries, making it one of the most important ports of China. This city was one of the original treaty ports; it also had a significant emigration during the 19th century, especially to southeast Asia. Most natives of Fujian province, including those from Xiamen county, emigrated to the Philippines, Singapore and Indonesia, with a smaller proportion of them heading to the United States and Japan. During the first period of China's opening to the international community in 1880s, a significant European community settled in the concession on Gulang Yu island. This same island later became the home of many wealthy overseas Chinese, who returned to Amoy to build their lavish dwellings. The island today is a curious reminder of that past, and constitutes an important tourist site on the South China coast. The links to the overseas Chinese community are particularly strong, and much of the investment that is planned from abroad will come from that source.

Xiamen consists of two islands and a mainland territory situated in a bay with a good natural port.

The traditional industrial structure was light industry: light electronic goods, some textiles, machine parts, and processed food made up much of the industrial output. The city of Xiamen is the second most important port in the province, and carries on brisk international trade, utilising its overseas Chinese connection for marketing abroad. In 1983, Xiamen earned more than $120 million through exports, accounting for a large portion of provincial earnings.

Although the industrial base is small[13], the factories and machinery are not modern, and significant investment is necessary to up-grade existing capacity for both export and selective import substitution. It is in the light of these new policies introduced in China in 1979 that the central government decided to give Xiamen an extended role in helping motor the modernisation of the country[14].

The Xiamen SEZ was established in October 1980, through the usual channels of central government and provincial government approval. The plan for development called for a rapid increase in investment, to be centred on light industrial build-up in the Huli district. Huli was designated an "industrial park" and, initially, the SEZ was limited to the Huli district. However, early in 1981, officials of the SEZ made it clear that they were willing to extend the advantages of the SEZ to the whole of the island of Xiamen, an important concession to foreign residents who were previously not allowed to live outside SEZ boundaries. By 1984, such residence restrictions were considerably relaxed. In April 1984, the territory of the SEZ was extended to the whole of Xiamen City (131 sq. km.), as well as Gulang Yu Island. This was the result of intensive lobbying by local authorities to create more attractive conditions for foreigners who wished to invest in hotel or tourist facilities, which could not be located inside the industrial district. There was even a hint that the conditions offered to the Xiamen zone may be extended to the whole of Fujian province, something that would virtually open the province to foreign residence and investment[15].

The chief obstacle to export-oriented industries was the lack of sufficient port facilities and poorly maintained municipal infrastructure.

The authorities of the SEZ began immediately to funnel Chinese state investment into the renewal of obsolete facilities and the building of new port berths. Overseas Chinese investment was also attracted to help construct warehouses in Huli.

The Huli district is on Xiamen island, some 7 kilometres away from the city centre and 3 kilometres from the new large Dongdu port. It is 2.5 kilometres from the new international airport that has been opened in Xiamen[16].

Master Plan

According to the Master Plan, the SEZ was to be constructed in two phases. The Huli industrial district (2.5 sq. km.) was begun in 1981. About 1.4 sq. km. have been levelled for construction since 1981. Other infrastructural projects already completed in 1984 include four deep water berths for ocean going ships (at Dongdu Harbour, adjacent to Huli) and water and power supply networks. In the Huli district, new paved roads now stretch 11 223 metres, and five completed factories offer 89 279 sq. m. production space. Fourteen more factories were still under construction by the end of 1984. The SEZ has, in fact, not begun to function as an industrial export zone. In a second phase, (1985-1988), plans were made to expand factory space to some 200 installations[17].

The Huli district was set up to attract investment for new installations. The SEZ, however, also aimed at attracting foreign direct investment to renovate existing enterprises, and this "import substitution" strategy was something the SEZ shared with Shantou, which also has an established production base. Renovation projects have certain advantages for foreign investors, including the fact that the output of the factory is already introduced onto the domestic market, and the Chinese partner takes the responsibility for distribution.

Table 28.   XIAMEN BASIC DATA, 1984

Location:	Central China coast, facing Taiwan.	
Area:	1 510 sq. km. for whole district; (SEZ is 131 sq. km.)	
The economy:	Industries:	Power generation, electronics, machinery, shipbuilding, precision instruments and meters, chemicals, pharmaceuticals, textiles, food processing, plastics and leather.
	Agriculture:	Rice, peanuts, sugar cane, tea and fruit.
Foreign trade:	Export volume: $120 million; 81 per cent of exports are mineral products, and a large protion of canned goods.	
FDI:	153 contracts to end 1984, with overall pledged investment of $716 million. $405 million was foreign direct investment. Rising portion of United States and Japanese investment in first half 1984, during which more than half of investment volume was committed. 59 projects have been completed.	
	The Huli industrial district adjacent to Donghu harbour is 2.5 sq. km. It is destined to be the industrial park of the SEZ; managed by the Xiamen United Development Company, it will seek priority investment in: electronics, instruments, machinery, chemicals, building materials, textiles, food processing, fibres, and chemical and biological engineering.	

Source: Intertrade, November 1984.

Generally, these types of foreign investment agreements have been disguised forms of compensation trade in which the foreign partner contributes technology and machines as equity, and the Chinese partner pays profits in exportable goods.

The revised schedule for the first phase of the building called for 160 000 sq. m. of factory space and infrastructure, providing some 30 000 to 50 000 new jobs. The total foreign investment target was fixed at $300-400 million. This sum has already been pledged. Much of the new pledged initial investment was in real estate; however, more than one half of the 103 contracts in 1984, involved industrial production, mostly in the electronics sector[18]. It was likewise specified that larger industrial plants could be set up in the Malian district, opposite Huli on the mainland outside the SEZ[19].

Xiamen adopted similar investment incentives to those promulgated for Gunagdong province. In late 1984, Fujian authorities announced that they intended to issue special regulations for Xiamen SEZ, and that these would differ from those in force in other SEZs. Table 29 summarizes the existing incentives for Xiamen. Annex 3 summarizes the different incentive packages for other SEZs and compares them for China. Xiamen changed the personal income tax incentive for foreigners living and working in the SEZ in 1984, giving a 50 per cent tax cut on income earned in China. Special incentives were also promised to Taiwanese who brought foreign direct investment to Xiamen, and it was reported in 1984 that at least one Taiwan concern had begun discussions for direct investment in Xiamen.

The first foreign joint venture went into operation in late 1983 in Xiamen. A brick-making factory was set up with Singapore capital to produce tiles. Four million dollars

Table 29. INVESTMENT INCENTIVES

Category	Incentive
Equity	25 per cent minimum, 100 per cent maximum foreign part.
Profit tax	15 per cent for manufacturing industries.
Tax holiday	Exemption two to five years, from first profitable year, upon negotiation; depends upon cycle of capital turnover and capital invested, technology levels.
Import taxes	Machinery, spare parts, raw materials, vehicles and other means of production for enterprises inside SEZ are exempt from import duties.
Depreciation	Depreciation on equipment goods can be recuperated early in contractual period.
Remittances	Profits, after taxes can be remitted abroad through Bank of China or any authorised zone bank.
Land rent	Twenty to sixty-year leases average, but may run higher depending upon investment. Land use may be transferred.
Rent/rate	Rmb 1-3 per sq. m. plus "land development fee" (Rmb 60-500), depending upon location and whether previous occupants must be relocated.
Wages	Rmb 50-60 monthly plus bonus paid to worker. Cost to employer 150 Rmb including bonus, welfare fund health and medical services.

Source: As cited in the Economic Reporter, September 1981, "Xiamen SEZ Offers More Generous Terms", p. 16; and China Economic News, January 1985.

111

Table 30. EVOLUTION OF PLEDGED FOREIGN DIRECT INVESTMENT (CUMULATIVE)

Contracts	Type	Amount ($ million)	Year
11	Joint ventures, Co-Prod.	1	1980
N.a.	Joint ventures, Co-Prod.	10	1981
N.a.	Joint ventures, Co-Prod.	161	1982-83
153	Joint ventures, Co-Prod.	405	1984

Source: B. Louven, "Die Wirtschaftssonderzonen der Volksrepublik Entwicklung und Modernisierung". *China Aktuell*, November 1983 and *Intertrade*, November 1984.

were invested in the project to date. The production materials were sent by container from the Federal Republic of Germany, transferred to the factory and processed[20].

Table 30 summarizes the evolution of pledged foreign investment in Xiamen.

By the end of September 1984, there were twelve joint venture contracts signed for the SEZ of Xiamen. Most of these projects are still in the development stage, and are under construction. Among the joint ventures set up are: a tile factory (Singapore capital), a plywood factory (Philippines, 100 per cent owned), an electronics firm (Hong Kong joint venture), cigarette factory (United States joint venture), and aluminium products factory (Hong Kong), a jewellery factory (Philippines) and an aquatic products company (Hong Kong joint venture). Other joint venture agreements have not been disclosed. Conflicting figures for foreign direct investment in Xiamen come from the problem of reporting investment. Several Chinese sources have indicated that domestic Chinese investment, including other provincial direct investment, could be as high or higher than the figures quoted for the foreign direct investment.

A major consequence of the extension of the SEZ to the whole of Xiamen city was the reorganisation of the SEZ authority. Previously, an Administrative Commission dealt with the planning aspects, and the Xiamen Construction and Development Company handled the contract negotiation and site overseeing at Huli. In August 1984, the Administrative Commission was closed down, and many of its functions moved to the municipal government office. A second Development Company was created, the Xiamen United Development Company (XUDC). This company is an interesting new type of business venture in the SEZs. The XUDC is in fact a joint venture among the Xiamen Construction and Development Company, which holds 57 per cent of the equity, and five Hong Kong banks, part of the so-called "sister banks" of the Bank of China[21]. In this way, the Bank of China is directly involved in the development of Xiamen, and is able to channel foreign exchange reserves into projects. The new United Development Company has been put in charge of Huli district, thus giving it a similar role to that of the China Merchant Steam and Navigation Company in the Shekou industrial district of Shenzhen. The creation of a specialised company for industrial development is significant. Chinese authorities, impressed by the modern management techniques of the CMSN, have made a point of creating a development company with a Hong Kong connection (Zhuhai has done likewise with the creation of Everbright). Consulting firms affiliated with the United Company have also been created to advise Chinese and foreign investors about both renovation and new enterprise projects.

Xiamen is beginning to transform itself under the impact of foreign investment. Two new hotels have been negotiated, and a third renovated by a Hong Kong firm. A new 27-storey international financial building is under construction, and is scheduled for completion by

1987. A large residential estate was begun on a 30 sq. km. site to house foreign residents. The largest single project under construction is a HK$ 400 million joint venture between Xiamen and Hong Kong companies to build an international exhibition centre by 1989. The municipal government authorities in Beijing have invested Rmb 400 million in Xiamen to build a hotel and a series of restaurants for Chinese tourists.

Xiamen is moving toward the development of a large tourist base and light industrial export and import substitution bases, in keeping with its traditional role as a port city, open to the outside world. The designation of the Xiamen as a SEZ is the first – and most important – step in ensuring that foreign direct investment flows back into China through its privileged status for investors.

Shantou SEZ

The city of Shantou was an area of high emigration toward the Americas and southeast Asia during the 19th century. The port was also one of the original treaty port cities where trade and contact with Europeans and Americans was carried out after 1843.

Shantou was chosen as a SEZ because of its good port facilities, its extensive network of overseas Chinese and the traditional trade and commerce activities of the city.

Table 31.   SHANTOU BASIC DATA, 1984

Location:	South China coast, Guangond Province, 180 nautical miles north of Hong Kong.
Area:	256 sq. km. (*of which:* 52.6 sq. km. are the SEZ).
Population:	9 million in the administrative county of which 730 000 are in the district of Shantou; 440 000 in Shantou City.
The economy:	Industries: Photosensitive materials, electronics, textiles, canned goods, furniture, plastics, handicrafts and traditional goods.
	Agriculture: Fruits, vegetables, maritime products; plans call for developing a 19.4 sq. km. area to emphasize horticulture, animal husbandry, fruits, vegetables, and fisheries; 300 acres will be allocated to spices.
Foreign trade:	The district exports are mainly light industrial products, (electronic instruments), such as handicrafts, embroidered goods, and bamboo ware, canned goods, also light-sensitive chemicals.
FDI:	Fifty-nine projects signed with a total of Rmb 40 million (29 are in operation). The SEZ absorbed HK$ 44.06 million in foreign investment in 1983. The net income of operations totalled HK$ 11.27 million. The Longhu District (22.6 sq. km.) and the Guangao District (30 sq. km.) are industrial zones within the SEZ. The Longhu area wishes to establish 250 enterprises with a total employment of 50 000 employees, emphasizing food processing, electronics, transportation, telecommunications, tourism, agriculture, petrochemicals, daily-use chemicals, household appliances, textiles, plastics, metals, and handicrafts. The Guangao district will emphasize petrochemicals with a 50 000 ton capacity oil wharf and 20 000 ton oil wharf. Existing industries include 370 enterprises, of which 120 are engaged in export production.

113

Shantou has had a slower pace of development than the other zones. The initial plans called for developing tourism and export processing of light manufactured goods in the zone. Two areas were designated as part of the SEZ within the city limits of Shantou. The Longhu district (22.6 sq. km.) on the outskirts of Shantou proper is a hilly area, about 15 kilometres from the port facilities, and 3 kilometres from the city centre. A small airport is located about 17 kilometres from Longhu. A second area of 0.2 sq. km. was set up for light manufacturing industries and the major concern of planners was to limit the building of new facilities in the SEZ to low pollution enterprises and relatively energy-efficient installations.

Shantou wished to distinquish itself from the other SEZs from the beginning. Mr. Chen Xu-fan, vice-president of the SEZ management company, stated the objectives clearly: "The two other SEZs are engaged in multilateral developments ranging from property to tourism and industries, but ours will be focusing on a few lines such as processing work and transportation facilities". The original plan called for a development in two stages[22]. The first stage was avowedly labour-intensive, and required small- to medium-sized investment capital (up to HK$ 1 million). It would involve elementary processing work. The original plans called for export processing of textiles, ready-to-wear goods, electronic meters and instruments and household chemicals. This list was later modified by the authorities of the zone, who sought short-term investment in more specific product lines[23]. Most of these investment projects were renovated projects rather than new factories.

In a second stage, the SEZ was to attract capital-intensive industries (up to HK$ 8 million) and larger investment packages. This is also the aim of the other SEZs, all of which want to move from labour-intensive industries to capital-intensive as soon as possible. This has led to competition among the zones for available foreign investment sources.

As Shantou had an existing production structure (unlike Shenzhen and Zhuhai, for instance), foreign capital will principally be used to upgrade existing facilities. Light manufactured goods enterprises already exist in the chemical fibres, photographic materials, electrical appliances, electronic components, building materials and some glass goods.

In late 1981, the Party Committee of Shantou County decided to establish four different development corporations, including a commercial agricultural company (which also operates outside the SEZs), a tourist company, a service company and a development management company for the whole zone[24]. By the end of 1982, the general plan of development for Longhu had been made public[25]. This plan was ambitious in scope, probably due to the initial success of Shenzhen in attracting foreign investment.

By May 1983, eight contracts with foreign companies had been signed, ten projects were being discussed, and 23 more were subjects of feasibility studies. Two refurbishing contracts had been signed for Chinese factories located outside the zone. Of the eight contracts already signed, the majority were with Hong Kong firms. These contracts included agreements for 100 per cent foreign-ownership investment, co-operative investment with the Shantou SEZ

Table 32. SHANTOU FOREIGN DIRECT INVESTMENT

Firm	Type	Amount pledged	Production	Partner
Zheng Da	Rug factory	HK$ 4 million	15 000 sq. ft./yr.	100 % HK
Kangdi Zhengda	Animal food	Under neg.	7 200 tons/yr.	(US/HK)
Jin Tai	Decoration	n.a.	n.a.	SEZ/HK
Wen Taili	Commercial cen.	HK$ 2 million	n.a.	SEZ/HK
Jin Tai	Hotel	HK$ 2 million (loan)	n.a.	SEZ/HK
Shan Tai	Transport	HK$ 20 million	Ship 3 300 tons	SEZ/HK

Source: Jingji Daobao, Hong Kong, 1983.

Development Company, joint ventures between Hong Kong firms and foreign firms with Chinese partners, financial co-operation in the form of loans secured from foreign sources by Chinese contractors and compensation commerce[26]. Of the 23 feasibility studies, 20 were for industrial manufacturing plants, two for animal breeding facilities (pisciculture and mechanised chicken breeding) and a self-service restaurant. Potential investment partners include firms from Hong Kong (13 projects), Thailand (ten projects), Canada (six projects), Singapore (two projects) and France (one project). Table 32 gives an overview of investment as of June 1983 in Shantou.

Investment Performance

The six principal contracts that have been signed involve the various categories of agreements summarized above. Details of them bring out the rather sanguine response of foreign investors to Chinese appeals:

1. The US-Hong Kong joint venture Kangdi Zhengda will manage a factory of animal feed. The project is still at the stage of planning the setting up of the factory and the respective contributions of the different partners.
2. The international investment company Zheng Da (Hong Kong) is financing 100 per cent costs of a rug factory. This is a holding company of Kangdi Zhengda.
3. A co-operative joint venture will be set up between Jin Tai (Hong Kong) and the Shantou Development Company to establish a medium-sized interior decorating firm. The Hong Kong firm will supply capital, equipment and techniques, the SEZ company will supply land and buildings. Part of the production is for export, the firm will have the possibility of importing into the PRC as well.
4. A commercial centre will be set up in Longhu by the Wen Tai company of Hong Kong and the Travel Company of the Shantou SEZ. Hong Kong will furnish part of the capital, to a ceiling of HK$ 2 million. The SEZ Development Company will provide the buildings, the equipment and a part of the capital.
5. The Jin Tai company of Hong Kong will furnish a loan for HK$ 2 million to set up a hotel in Longhu. The loan will be repayable over four years.
6. The Shan Tai Maritime Transport Company has been created as a joint venture between the Hong Kong based Dazhong Travel Company and the Shantou SEZ Travel Company. This company will carry passengers between Shantou and Hong Kong. The capital will be provided by the Hong Kong based firm (HK$ 20 million).

The surprising element in this list is the number of firms involved. In fact, the Hong Kong based Zheng Da and the Jin Tai account for three-quarters of pledged foreign investment by the end of 1983. This is a sign that the Shantou SEZ has not been able to attract the diversity of foreign capital in the same way as Zhuhai or Shenzhen.

Shantou faces development problems due to its poor infrastructure. Although the port is well situated in Guangdong province, connections with the rest of China are poor. For this purpose, the Shantou County Government is prepared to invest in infrastructure as a means of attracting capital. The Chinese capital investment involved has not been officially published.

By the end of 1984, it was reported that 18 joint ventures and co-production enterprises had been set up in Shantou[27], although the exact breakdown of sectoral investment and foreign participation has not been given; the failure to distinguish between pledged and realised investment also makes it difficult to estimate the SEZ growth. The general lack of official data would suggest that investment performance has been poor. It was reported that all the joint ventures were with Hong Kong firms, principally in the area of plastics, toys, ready-to-wear garments, rugs and furniture.

From 1982 to 1984 the SEZ authorities constructed two hotels, a business centre, two large warehouses, four general industrial buildings, several new roads, as well as a new telephone installation and a new water supply system.

In addition to the above, the following items of infrastructure were announced in mid-1983 as underway: a self-service restaurant (2 400 sq. m.), a commercial centre (1 350 sq. m.) and a front office for the SEZ Development Company (200 sq. m.), and three new buildings for industrial installations (10 000 sq. m.).

Table 33. SHANTOU INFRASTRUCTURE: PLANS AND DEVELOPMENT, JUNE 1983

	Planned	Under construction	Completed
Roads	15.5 km.	3 km.	4.2 km.
Bridges	3	n.a.	1
Retaining wall	4 km.	n.a.	1.3 km.
Drains	42 km.	3 km.	3.6 km.
Water treatment	1 plant	n.a.	n.a.
Electric relay	1 station	n.a.	n.a.
Transformers	10 stations	n.a.	n.a.
Electric cables	56 km.	6 km.	6.6 km.
Telephone lines	1 900	n.a.	100
Tree plantation	16 800	–	16 800
Port berths	2 (5 000 tons)	2 berths	1 berth
Industrial space	200 000 sq.m.	100 000 sq.m.	23 760 sq. m.
(reserve)	1 000 000 sq.m.	–	–
Housing, commercial	56 000 sq.m.	n.a.	6 160 sq.m.

Source: Jinqji Daobao, op. cit.; Investment Guide, SEZ Shantou, May 1983, (in Chinese).

It is apparent, from the size of infrastructural projects, that Chinese capital investment – as in the other three zones – far surpasses foreign investment flows. It is the hope of Chinese authorities that the first stage of SEZ build-up will reap benefits, once the proper investment environment has been established.

The incentive package that was published in April 1983 is the most complete statement to date of investment incentives. It resembles packages offered for Shenzhen and Zhuhai; these latter are compared in Annex 3. Shantou, which does not have the advantageous position of Shenzhen and Zhuhai, has added more flexibility to its incentive package, offering longer periods of rental or lower rates on capital-intensive industries[28]. Shantou has generally attracted Hong Kong investors wishing to relocate for cost purposes or interested in investing in local renovation projects that will later afford them access to the domestic market for production within the SEZ. In this sense, the SEZ is following a policy of renewing existing enterprises with foreign capital, rather than trying to attract new enterprises. The reported number of joint ventures (between six and eighteen, depending upon the sources) have not been officially verified by SEZ authorities, and caution must be used in interpreting the data provided for Shantou.

Shantou Area Manpower Pools

The Shantou area (city and surrounding county) claims to have over 100 000 youths between the age of sixteen and twenty-five, ready and able to work. Each year, this number is augmented by 7 000 new arrivals on the work market. At least 70 per cent of this population has graduated from primary school. This work force can only be absorbed by the creation of new industry in the area. Some 2 000 technicians live in the area and in the range of 740 engineers; this includes "high" qualified engineers, plus both "ordinary engineers" and assistants to engineers; these groups differ according to education, the "high" having received a college education, the assistants have received para-professional training after simplified high school studies. In addition to this specialised manpower pool, a large pool of skilled workers exists (industrial chemical works, electronics, precision instruments, plastics, textile and artisian workers). The Shantou area has a specialised technical school for the formation of workers in the SEZ itself. Wage levels for workers were set at HK$ 400-600 in 1982[29]. By mid-1983, wage levels had risen to an average of HK$ 1 000[30].

Investment Incentives

Among the special conditions accorded to foreign investors are the long-term leasing arrangements. These differ with the type of contract and the amount of money invested. A considerable degree of latitude was granted to foreign investors who agreed to set up import-substitution industrial operations in the SEZ rather than concentrate on processing or assembly type operations. As an example of structural investment incentives offered to foreigners, Table 34 gives the conditions for the land lease in Shantou.

Table 34.   LAND LEASE CONDITIONS: SEZ SHANTOU

Activity	Ground rents[1]	Lease time (years)
Real estate	Rmb 14-40 sq. m./yr.	50
Education, health	"Preferential"	50
Industrial	Rmb 5-20 sq. m./yr.	30
Tourist projects	Rmb 30-70 sq. m./yr.	30
Agriculture	Case-by-case	20
Commercial	Rmb 35-140 sq. m./yr.	20

1. Varies according to level of investment and location.
Source: Investment Guide, SEZ Shantou, May 1983 (in Chinese).

Industrial investment is the most favoured in rent terms, but real estate has more favoured lease time, an inducement to investors to see returns on investment over a sufficiently long period of time. Industrial investment would expect returns on investment in less time.

The planning authorities of the SEZ have made provisions for granting high technology installations free rent if the enterprises are judged to be "essential" to the national economy. Other provisions include stipulations for revising land rents every three years; in no case can the rate of increase exceed 30 per cent of the previous level of rent. It is the general rule that rents should be paid in foreign currency or Foreign Exchange Certificates by joint venture firms.

The land lease situation in Shantou – as in the other SEZs – is one of the principal complaints of foreign investors. This investment "disincentive" is discussed in Chapter 5.

Other incentives created to attract foreign investors include the usual 15 per cent tax on profits obtained within the zone. The Shantou investment manual also makes it clear that the tax rate may be negotiated downward with a foreign investor under special conditions, which include the use of foreign capital over a fixed sum (this sum is not specified in the *Guide*), and the setting up of joint ventures which involves the use of "high" technology. This latter term poses special problems, as it is technology as defined by the Chinese partner.

The Investment Guide also mentions conditions under which goods produced within the SEZ can be sold on the domestic Chinese market – a situation that has not been obtained until very recently[31]. This access to the domestic market is clearly to be treated as an incentive.

The Chinese permit access to the domestic market if the essential input in raw materials or in individual parts is of Chinese origin (therefore, imported from China, and subject in certain cases to duty when entering the SEZ). In no case can the constituant parts be imported from abroad. If the joint venture attracts a sufficiently high enough level of foreign investment (the exact figure is not quoted) then the enterprise may engage in domestic sales. Lastly, if the SEZ product is manufactured with "high technology" (as defined by the Chinese), then the domestic market is also open for sales[32].

However, until the convertibility of the renminbi is decided, SEZ firms engaging in domestic sales would have to accept renminbi, or make special arrangements to sell their product for Foreign Exchange Certificates only (convertible renminbi). This latter would limit severely the potential domestic buyers, as FEC are only available as exchange certificates for convertible currencies. A separate section of the study will deal with the problem of opening the domestic market to products from the SEZs.

Shenzhen Special Economic Zone

Shenzhen SEZ is the most developed of the four SEZs. It is also the most studied in recent literature[33]. Its location, adjacent to the New Territories of Hong Kong, makes it an ideal location for Hong Kong investment. It also profits from the large external economy of Hong Kong. The Shenzhen zone was the first to have actual foreign investment. It was, in some ways, the pilot project of the SEZs.

The provincial laws that established the SEZs in Guangdong Province are the genesis of the transformation of Shenzhen[34]. Prior to March 1979, Shenzhen was a county (Bao'an xian) in the administrative hierarchy. In this way, it depended upon the Prefecture of Huiyang for decisions concerning its development. The first step in granting special status to the Shenzhen area, was to change its administrative structure. The county was divided, and Shenzhen was raised to the rank of a municipality (shi) which depended directly upon the provincial government[35]. Within the newly created municipality of Shenzhen, a special area was set aside for the SEZ. Shenzhen occupies 327.5 sq. km., or about 16 per cent of the new municipality's territory. This area was decided upon after the idea of making the whole municipality a SEZ was abandoned in 1980. By 1984, the whole of Bao'an county had been reorganised to include the municipality of Shenzhen and the SEZ. The administrative centre for this new municipality unit (1 700 sq. km.) is to be the municipal government of Shenzhen, thus extending the power of the Shenzhen city government even further.

Table 35. SHENZHEN BASIC DATA 1984

Location:	Behind Hong Kong, along Guangdong Province border.
Area:	327.5 sq. km. (*of which* 30 sq. km. urban).
Population:	300 000 (*of which* 100 000 temporary construction workers).
The economy:	Industries: Electronics, textiles, oil, petrochemicals, footwear, steel shipping containers. Of the Gross Value of Industry (Rmb 802 million in 1983), over one-third comes from foreign enterprises.
	Agriculture: Dairy products, poultry and vegetables.
Foreign trade:	Hong Kong absorbs more than 70 per cent of official trade, which is in part trans-shipped internationally. Unofficially, much trading goes on with the rest of China.
FDI:	By end June 1984, about 3 272 agreements for a value of $1.8 billion. Priorities for industry: electronics, petrochemicals, oil, food, textiles, construction, materials, precision instruments, tourism, agriculture, animal husbandry, building supply base for South China Sea oil exploration and development. Large real estate investment, principally from Hong Kong (80 per cent of investment).
	The Shekou industrial District (2.14 sq. km.) is part of the Shenzhen SEZ, but is managed by the China Merchant Steam and Navigation Company, based in Hong Kong. The priorities for Shekou are: yachts, marine containers, structural steel, aluminium, electronics, toys, food, furniture and tourism.

Source: Intertrade, November 1984.

One of the peculiarities of the new Shenzhen SEZ was its multipurpose goals. Unlike other export processing zones with largely industrial operations, or even other SEZs in China which advertised industrial and service sector goals, Shenzhen set out to be an enclave of fully integrated, market-type sectors. This multisectoral goal is in fact largely possible through the planned complementarity with the Hong Kong economy.

Initially, the SEZ was to be divided into three type of investment: real estate, tourism and industry. The industrial zone of Shekou, administered by the China Merchant Steam and Navigation Company has special industrial projects, and is, in some ways, a competitor for FDI within the SEZ. In order to administer the SEZ, the Shenzhen Development Company was set up to act as SEZ agent in negotiating land leases, and receiving foreign investors. Parallel to the company, there is a municipal government, headed by a mayor; the first mayor, Liang Xiang, is also vice-governor of the Guangdong Province, thus assuring a high level of contact for the SEZ in the provincial government. As the bureaucratic problems posed by the creation of the SEZs are not unimportant, it is interesting to note that this level of authority was assured from the inception of the SEZs.

In 1983, a special office was set up in the municipal government to deal with attracting investment to the zone, leaving real estate negotiations to the Shenzhen Development Company. This coincided with the change of policy in the SEZ.

The authorities of the zone announced that Shenzhen wanted more "productive investment" (manufacturing and industry) and less real estate and tourist investment.

119

The industrial evolution of Shenzhen was perhaps the ambition test of the SEZs. In opening the SEZs, the central government acquiesced to the argument that these special areas in the south might be able to attract new industries and bring with them new forms of advanced technology, even if it were only in the light industrial manufacturing sector. In the case of Shenzhen, this was a difficult challenge. Hong Kong industries, largely structured into light consumer manufacturing, were serious competition for Shenzhen. Hong Kong had a better labour force, a much better infrastructure, and although wages were higher, so too were profits[36].

Of the 327.5 sq. km. originally allocated for the SEZ, 15 were targeted for industry. The sectors to be developed in the SEZ were very similar to Hong Kong industrial installations: electronics, textiles, food preserving, furniture, machine goods, building materials, petro-chemical goods. These were developed in ten sites[37]. Six industrial companies were also set up to handle business ventures; a liaison office, the Industrial Service Development Company handles negotiations, supply and the setting up of the industrial site. The latter also serves as a link to the specialised companies.

The conditions offered to foreign investors were similar to those offered in the other zones: a negotiable 15 per cent tax rate, three to five-year tax holiday for specialised industry, import duty reductions on materials used for production, flexibility in hiring and firing, remittance of profits abroad, income tax dispensations for foreign workers, etc. (See Annex 3 for complete list of incentives.)

The obvious first step was the construction of the necessary infrastructure. Shenzhen in 1978 was little more than a small-sized town with a customs and train station, and an elementary road system; Shekou zone was virtually empty land. To launch the SEZ, the Provincial government was obliged to invest in infrastructure and basic facilities in order to attract initial investment. In a recent article, one scholar has estimated the cost of Shenzhen infrastructure investment to Chinese authorities to be Rmb 2 000 million for the period 1984-2000. This figure is certainly too low[38]. In this regard, the Shekou area of Shenzhen, administered by a semi-commercial firm based in Hong Kong, the China Merchant Steam and Navigation Company had a decided advantage over the Shenzhen Government, as it was able to raise funds from outside the country without approval of the government; as the Shekou zone is small in size, infrastructure was initially less expensive. Shekou has had the advantage of being a potential port, and manufactured goods could be shipped out to Hong Kong without going through Shenzhen, thus ensuring foreign exchange earnings to repay loans. In the SEZ in general, foreign investors have been asked to finance some of the infrastructural costs themselves; this has slowed down development (sometimes this is done indirectly through "land use fees"). There were a number of modifications to the master plan in the first year of operations (1980), and several roads that were nearing completion were abandoned. The central road linking Shenzhen city and Shekou directly is at times a two-lane highway, and will have to be modernized soon to ensure a free flow of traffic within the SEZ. Other new roads will also have to be built to permit access to residential areas and railway supply stations. Thirty-seven new roads have been completed in the zone totalling 56 kilometres. Shenzhen now has 80 kilometres of paved roads with good links to the major road connecting Guangzhou and Hong Kong. Plans called for a four-lane highway to link the zone to Guangzhou and to Gongbei near Macao[39]. There are rail links to Hong Kong and to Guangzhou, and a new station is being built to accommodate the estimated 35 million passengers a year in the year 2000 (1983: 8 million passengers). Double tracking is being planned for the Guangzhou-Hong Kong line, and a small SEZ internal railway will also be built. There are currently eight sea routes to Shenzhen linking it directly to Hong Kong, Guangzhou, Shanghai, Dalian, Lianyungang, Qingdao, Tianjin, and Zhanjiang. Berth

Table 36. LAND USE PLAN FOR SHENZHEN SEZ

Parts	Section	Districts	Major functions	Usable land (hectares)
Eastern part		Dameisha	Tourism	
		Xiaomeisha	Tourism	1.72
		Yantian	Fishery, industry, agriculture	5.78
		Shatoujiao	Commerce, residence	2.60
		Liantang	Industry	3.00
	Eastern section	Luohu City	Commerce, industry, residence	2.00
		Old City	Commerce, industry, residence	4.00
Central part		Reservoir Dis.	Tourism, residence	4.40
	Central section	Shangbu Dis.	Industry, residence, warehouses	10.00
		Futian New urban area	Comprehensive	30.00
		Chegongmiao	Comprehensive	6.00
	Western section	Xiangmihu	Tourism	2.10
		Agronomic Ins.	Scientific research	
		Garden for Scientific Research		4.00
		Shahe Dis.	Mainly industry	12.00
		Shekou Dis.	Mainly industry	2.30
Western part		Houhai Dis.	Comprehensive, education	6.00
		Xili Reservoir	Tourism	3.00
		Chiwan Port Oil Base	Industry, port, comprehensive	5.00
		Nantou, Baon county	Industry, comprehensive	6.10

Source: Investment Guide for Shenzhen Special Economic Zone.

facilities exist at Shekou (3 000 DWT with a draft of more than 5 metres), and a daily hovercraft links Shekou to Hong Kong. In 1984, construction began on Chiwan port, south of Shekou. This port will provide facilities for 10 000 DWT ships, while the third port of the zone, Mawan will accommodate 50-100 000 DWT vessels. Authorities are hoping that by providing infrastructure for oil tankers, they will be able to attract service bases for the South China Sea oil companies, thus replacing Zhanjiang, which is neither well placed geographically nor entitled to SEZ investment incentives for foreign investments. A container terminal is being built at Yantian and harbours are being studied for the Dongtoujiao oil depot area and the tourism installations in Xiaomeisha and Dameisha. A helicopter service is now operational between Shenzhen and Zhanjiang, the oil base in Guangdong.

The SEZ infrastructure includes a new barrier fence of 88 kilometres long that will separate the SEZ from the rest of Boa'an county. This demarcation line will in fact constitute a border with the PRC once finished and put into service. SEZ authorities plan to seal it and simultaneously relax control on the Hong Kong-Shenzhen border, which now has three passages points, and at the same time ease customs and immigration regulations. Luohu is used for passenger traffic (on the main Hong Kong-Guangzhou line), Wenjindu checkpoint for vehicles and Shekou for freight and passengers arriving by sea. All three have customs control.

The FDI real estate investment has been by far the most important in absolute terms (68.5 per cent). The majority of the contract agreements have been with Hong Kong investors, anxious to build in the SEZ with an eye to the settlement of the Hong Kong situation. The majority of these investments are commercial buildings and residential apartments which are destined for the relatives of Hong Kong families or rent properties intended for foreign businessmen. Two large hotels are planned for the zone, the Nanhai in Shekou ($10.9 million) and the Shanghai in Shenzhen ($3.8 million). In some respects, this property development is what the Chinese authorities have called "non-productive" or "non-value added" and they are making every effort to discourage this type of investment in the future. It is, in some sense, a way in which the Hong Kong investment market can move into the future "suburb" of Shenzhen.

Tourism is the third area targeted for direct foreign investment. The Shenzhen area will be used principally to build leisure facilities for Hong Kong residents, where recreation land is at a premium, and facilities short. Thirteen projects are currently under contract, including recreation and holiday facilities along the east coast of Shenzhen (among them the large Club Méditerranée installation at Dameisha for an estimated HK$ 20 million). These facilities are not, for the moment, available to residents of the PRC, as they are located within the SEZ, where special permission is necessary to travel. They are designed to attract weekend clientele and foreign tourists, including Japanese and Western tourists visiting Hong Kong.

Shekou Industrial Zone

The Shekou industrial district of the Shenzhen SEZ is of particular interest[40]. The one sq. km. area at the west end of the Shenzhen SEZ is under the direct management of the China Merchant Steam and Navigation Company (CMSN)[41]. It is also co-owned by the PRC Ministry of Communications which has representatives on its Board. This gives the Shekou industrial zone special access to high-level policy makers within the PRC; perhaps more importantly, the company has a good management base with Western experience. The results in Shekou reflect these two decisive advantages.

The CMSN, operating as a joint venture partner with foreign firms, began work in Shekou in 1979. The CMSN set about to construct the necessary infrastructure. By the end of 1983, more than 100 sq. km. of land had been levelled to prepare the area for industrial construction. Shekou also has a water front, a 600 metre dock was constructed to berth 5 000 ton ships (6 metre draught). The Chiwan district, south of Shekou, was also designated as a development site, and a harbour was built there for deep water ships of the 10 000 ton class. An 8.5 metre wide 8 kilometre long highway links Shekou to the Guangzhou-Shenzhen highway at Nantou, and to Shenzhen city, and via the latter to Hong Kong[42]. The nearby Xili reservoir provides a water supply (20 000 cubic metres a day). The power source for the Shekou area, like that of Shenzhen in general, comes from the China Light and Power Company (via the Hong Kong-Guangzhou power grid). The daily power supply is in the range of 1.35 million kwh. There remains a problem of maintaining a constant level of electric current, something that has not been resolved. Telecommunications are modern with 1 000 electronic programme controlled channels, and 600 microwave channels; there are also telex links to Hong Kong and international telecommunications via the Shenzhen exchange station.

More than 60 housing blocks were built for workers, 30 villas for foreign executives and their families, as well as seven apartment buildings where title to living space may be bought and transferred. Hotels, bars, restaurants and service centres were built. The new Shenzhen University was transferred to Shekou in 1984. An entire city was constructed between 1979 and 1983. Factory space was constructed for more than 20 medium-scale factories by the end of 1984; 30 were functioning, producing more than 40 varieties of light industrial goods, including furniture, toys, elastic bands, colour television sets, radio-cassettes, electronic calculators. Heavy industry is also represented by factories for rolled steel, aluminium sheeting and cargo container construction, yacht construction and paint products. Joint ventures in the Shekou district, which numbered more than 60 in June 1984, imported advanced production technology to the zone; 67 sets of advanced equipment were imported by 13 of the joint ventures, among which three were advanced world level technology sets, 37 advanced for China and four at a middle technological level for China.

The CMSN was especially interested in attracting industrial projects, and acting as a joint venture party with foreign business firms. Shekou authorities targeted "technologically-advanced, capital-intensive" projects, and favoured their development over classic labour-intensive industries.

Three types of enterprise operate in Shekou: 100 per cent CMSN-owned and managed firms; 100 per cent foreign-owned firms; joint ventures and co-production firms. The latest statistics indicate that more than 100 firms have signed contracts, and that at least 60 are in operation.

By March 1984, the total pledged investment from both the Chinese and foreign partners in Shekou totalled $200 million. The following table gives a breakdown of the investment inside the Shekou zone.

Table 37.   SHEKOU INDUSTRIAL ZONE INVESTMENT, DECEMBER 1983

Sector	Contracts	Operational
Industrial	38	23
Transport	3	3
Real estate	11	10
Hotel/Restaurant	8	
Commercial	16	14
Other	8	1
Total	84	58

Source: Interview of author with CMSN officials, December 1983.

The wide variety of investment projects reflects the interest of the CMSN to diversify its activities and acquire business experience outside the transport sector.

It was announced at the beginning of the Shekou operation that the industrial zone would not seek, as other SEZs would, certain types of production; the list itself is an interesting commentary on the way in which the company planned to specialise from the outset. The CMSN rejected forms of investment that were meant to transfer labour-intensive production units from Hong Kong to Shekou, or out-moded forms of production that were becoming increasingly costly in other developing countries of the region (in Singapore, and the Philippines for example). There was also an increasing consciousness that technology transfer was not simply a matter of using new machinery for production tasks; the machinery employed needed to be the latest generation of equipment and it had to be feasible in the production context of the Shekou zone. The five "unwanted" forms of foreign investment were enumerated in the official study[43]:

-   assembly work of outmoded machinery;
-   compensation trade;

- industries that were specially energy hungry, or posed pollution problems for the environment;
- machine piece production;
- production of goods that are traditionally exported from the PRC (especially textiles).

In a large measure, the CMSN has been successful in attracting investment in the desired categories, as the list of industries provided by the CMSM clearly shows.

Table 38. SHEKOU: TYPE OF INVESTMENTS/AMOUNT, DECEMBER 1983

Type of contract	Number	Amount
100 per cent CMSN-owned firms	10	n.a.
Joint venture-CMSN-foreign[1]	14	n.a.
Joint-venture-foreign-foreign	63	n.a.
Total	87	$200 million

1. Joint venture is taken to mean equity joint venture and co-production contracts.
Source: Interview of author with CMSN officials, December 1983.

Riding on the wave of official government support for private business and the "separation of business from politics", authorities in the Shekou area were able to attract more than $150 million of realised investment. Seventy per cent of that capital came from Hong Kong, and of the 65 contracts signed, 15 per cent of these were 100 per cent foreign owned. This is a higher percentage than elsewhere in the SEZs of China.

The CMSN has been accorded the entire financial responsibility for the development of the Shekou zone; this gave it a status analogous to that of a development company in the other SEZs. It had considerable advantages, however, over the regular development companies elsewhere in the SEZs, as has been mentioned above. There is good reason to believe, however, that the CMSN may be preparing to play another role in the modernization of China, and therefore, a closer look at the activities in the Shekou zone will be useful.

In an official article reviewing the performance of the Shekou zone in 1982, Chinese authorities spelled out both the expectations for the industrial zone and the latitude afforded the CMSN to accomplish those goals[44]. It was clearly apparent from this presentation that the Chinese authorities were interested in using the unique position of the CMSN to forward the rapid development of Shekou; it is equally apparent that the CMSN was seen as a potential Chinese multinational corporation that needed more experience in managing multifaceted economic activities in the controlled context of a Special Economic Zone. For that reason, the company was given special powers to conduct business and manage the zone with little, or no reference to the existing hierarchies inside the PRC. Before the creation of the SEZs in 1978, the CMSN had already begun to diversify its activity, which until then had been limited to shipping and industrial production in the area of transport. It began to finance projects from Hong Kong for supermarkets, restaurants, maritime supply stores and repair facilities for boats. This desire to "better its competitive advantages by enlargening its horizons" was part of a decision to internationalise the company and develop it into a larger firm with diversified investments and interests. As a private company controlled by Beijing, it would be the unique capitalist business arm of the Beijing Government. However, the principal drawback to this development was at once the principal advantage of the CMSN: it was located in Hong Kong, where competition for market shares was fierce, and any entry into a new sector would involve taking business away from existing firms, already well implanted.

The creation of the SEZ in Shenzhen gave the CMSN Company the comparative advantage they had been seeking in developing new horizons: they would be allowed to take over the implementation of an industrial zone, part of the SEZ, and then diversify themselves into all branches of economic activity related to the development of a large industrial park. By giving the CMSN full and monopolistic control over the Shekou zone, the Chinese Government endowed them with the protective business environment that would ensure their growth into new sectors. This growth, of course, is carried out by contracting joint venture agreements with foreign businesses, thus allowing the most widely experienced company under Chinese control to negotiate and set up industries that would permit the transfer of both management skills and the new technologies brought in by foreign business.

CMSN is obviously also competing for limited foreign funds with the Shenzhen Development Company, wishing to attract investment to the Shekou area instead of other designated Shenzhen industrial areas (such as Bakualing and Shangbu). The CMSN did not hesitate to put forward the advantages that it offered to foreign investors, employing the unique opportunities in Shekou[45]:

- the hierarchy of decision for investment approval is simple. The Chinese partner is the CMSN, which has direct access to ministerial level approval in the PRC and no intermediary bureaucracy in the SEZ or the Province of Guangdong;
- the CMSN is a totally autonomous body within the SEZ, and can sign contracts without referring to the Municipality of Shenzhen, something that has slowed down other foreign investment;
- the long business experience of the company gives them an advantage, as they are used to dealing with foreign expectations, and legal requirements, and have a sense of negotiation that is relatively unusual in the PRC;
- having a head office in Hong Kong, the company is able to follow political and economic trends in that city and secure loans on the open market in Hong Kong. It also has access to important information concerning international business cycles, something that few Chinese institutions have at the present time.

The $100 million invested by the CMSN in infrastructure for the zone was raised on the international capital markets without direct help from the Beijing Government, something that was seen as part of "doing it the capitalist way". The investments that have been made to date in the zone reflect the desire of the CMSN to attract new industrial ventures: by the end of 1983, 23 factories were in operation, representing a cumulative Chinese foreign investment of more than HK$ 1 billion (of which HK$ 600 million is Hong Kong investment, principally in manufacturing industries), and another 20 factories were under construction or in the planning stage[46]. However, one must be cautious with these statistics as the term "factory" is often used for what in Hong Kong is called a "flat factory"; this latter is basically a floor of a building, often an old apartment building, that has been converted into a production space. In Hong Kong, this practice developed due to the very high land costs; it is not considered to be an efficient means of running a factory, and the Hong Kong business community has been critical of the CMSN for having chosen this form of production unit for the Shekou zone, when land space was precisely what the new development had to offer. It was considered to be a sign that the Chinese had not done sufficient cost-benefit analysis of the situation in Hong Kong before embarking on work in Shekou.

The Shekou experience is seen by the Chinese authorities as a vanguard of the Four Modernizations Programme, and if a wide range of liberties have been given to the CMSN to develop the industrial district, it was as an experimental laboratory for the rest of China and much importance was placed upon the results obtained, as was clear from official declarations in 1982[47]. These statements were all the more surprising in that they indicate a certain critical attitude towards economic planning practices within the PRC, and more specially towards the management techniques applied to the state-owned enterprises. This of course reflects national policies, as is discussed elsewhere.

The CMSN has also been active in trying to attract offshore oil exploration firms with the promise of good service facilities in Shekou. Recently, it was decided to build a large deep water port at Chiwan (to accommodate ships of up to 10 000 tons). The infrastructure for Chiwan, which is located a few kilometres to the south of Shekou, is now under construction. This port is designed to become one of the oil bases in the SEZ. It was estimated that the cost of construction in Chiwan will be in the range of HK$ 300 million, to be raised independently by the CMSN. For the SEZs Chiwan is one of the three planned bases (with Zhuhai, Shantou and Mawan). The new port area is being set up and managed by the China Nanshan Development Company[48], in which the CMSN has the majority interest.

To the north of the Shekou zone and on the other side of the short peninsula that separates it from the open sea, the oil base of Mawan is also planned. The Shenzhen Petrochemical Industry Corporation was created (1983) to manage the Mawan installation. The first construction will be a petrochemical complex, designed to refine crude oil supplied from the South China Sea oilfields. The proposed order of construction will involve building a ship terminal for tankers (this will also mark the line between the Chiwan and Mawan zones), followed by a railway line to Shenzhen city centre. The thermal station for electricity generation (250 mgw. to 500 mgw.) will come in a second stage, as well as a refinery that will handle up to 5 million tons of crude. Once the area is in operation, production units for plastic derivative products will be built. There has been considerable foreign interest in this area of Shenzhen SEZ, as evidenced by the letter of intent signed by an Australian group (five companies in industry and engineering) for $500 million.

Daya Bay

The future site of the Guangzhou nuclear power station is located in the SEZ of Shenzhen (it is located some 60 kilometres from the centre of Shenzhen city on the coast). On 80 hectares of land, a generator is being planned in conjunction with the Office of the National Nuclear Commission, located in Shekou. This project is to be a joint venture with French, British and Hong Kong firms who will provide the technology for the nuclear plant[49].

NOTES AND REFERENCES

1. See *China Economic News,* (Hong Kong), No. 2, 1985.
2. See SWB FE/W 1333/A/6 "Perspectives for Pearl River Development Zone".
3. *Economic Reporter,* Hong Kong, June 1982, p. 19. See the same issue for details of development plan.
4. See "Results Obtained in Foreign Investment in the SEZs of Shenzhen and Zhuhai", Zeng Muye, *Gangao jingji* (The Economies of Hong Kong and Macao), Guangzhou, 1982. By 1984 32 hotels had been built to accommodate tourists and foreign residents.
5. According to statistics given by *Intertrade,* December 1984.
6. Cf. "The Zhuhai SEZ is excellent", in *Renmin Ribao,* 28th January 1985; and a somewhat confusing report on Zhuhai in *China Economic News,* No. 15, 1985, p. 8.
7. Present freight handling in Zhuhai SEZ is 250 000 tons. The Jiuzhou area of the SEZ can berth 1 000 ton ships, and the plan calls for four berths for both passenger and freight vessels, as well as two berths for 10 000 ton vessels.
8. The work programme for the harbour and oil base was reported in *Financial Times,* 7th March 1984.
9. See "Zhuhai's Hong Kong Connection", M. Ross, *China Business Review,* November-December 1984, p. 39 ff.
10. See Ross, *op. cit.,* p. 40.
11. A feasibility study for industrial potential in the region was conducted in 1984 by Morrison-Knudsen International, with the assistance provided by the US Government's Trade and Development Programme. See Ross, *op. cit.,* p. 40
12. For Amoy's historical note, see Fairbank, *op. cit.,* Chapter 1.
13. Xiamen currently has more than 770 small- to medium-sized industries, employing more than 100 000 workers. Light industry accounts for 70 per cent of total industrial output.
14. For a complete physical description of the SEZ, see the *Investment Guide to Xiamen SEZ,* produced by the Hong Kong and Shanghai Banking Corporation and the Construction and Development Corporation of Xiamen, Hong Kong, 1984.
15. See the statements of Xiang Nan, Secretary of Fujian Provincial Communist Party Committee in "Fujian Offers Incentives to Investors", *Economic Reporter,* 19th November 1984. (Hong Kong).
16. The international airport was financed by a loan from Kuwait (Economic Development Foundation). It opened on 22nd October 1983, with flights to Beijing, Shanghai, Fuzhou and Guangzhou planned. *Economic Reporter,* November 1983. The total cost of the renovation was $40 million.
17. See *The Retrospect and Outlook on the Foreign Economic Relations and Trade with Fujian Province,* (The Commission for Foreign Economic Relations and Trade), Fujian, PRC, July 1983 (In English and Chinese).
18. Cf. *SWB* FE/W.1321/A/15, 16th January 1985.
19. See *Economic Reporter,* September 1981, "Xiamen SEZ Offers More Generous Terms", p. 11.

20. Ding Quilan, "The City of Xiamen Under Development", *China Reconstructs*, August 1983, pp. 12-14.

21. The partners are: The Xiamen SEZ Construction and Development Corporation and the Trust and Consultants Corporation under the Bank of China, the Chiyu Banking Corporation Ltd., the Nanyang Commercial Bank Ltd., the Po Sang Bank Ltd., the Hua Chiao Commercial Bank Ltd., the Nan Tung Trust and Investment Corporation Ltd.

22. *South China Morning Post*, 23rd August 1983.

23. The list published in *China Market*, May 1983 included light industrial products such as colour photo paper, film, paper, cigarettes, special plastic products, nylon yarn, batteries; textile products such as yarn, towels, woollen cloth and fabrics, medium-grade and high-grade western style clothing; various electrical goods, such as computer memories, computer-controlled telephones, large- and medium-sized integrated circuits, electronic watches, electronic toys, cameras, a wide variety of electronic goods, food stuffs (mushrooms, soft drinks, biscuits); machinery such as electric motors, air-conditioners and freezers, as well as metals, wood and wood-steel furniture, glass and ceramics for building, some cosmetic goods, including luxury goods for Western markets. Some of these facilities already exist in Shantou, and the foreign capital is invited simply to up-grade existing installations. The precise nature of this list, published in a major trade journal, indicates that the Shantou authorities have probably identified some potential investment sources for these types of activities.

24. As announced on Radio Guangzhou, 9th September 1981 and reported in *SWB*, 9th September 1981. The Development Company has under its responsibility an import-export firm, a trust and investment company, a reality company and a united navigation company. "Recent Developments in Shantou Special Economic Zone", Huang Biaoxing, *China Market*, No. 10, 1982, p. 75.

25. As reported in *Wen Wei Po*, 1st February 1983, and in *SWB*, 23rd February 1983, the plan called for a first phase development of 0.55 sq. km. and in a second phase 1.05 sq. km.. Some 250 plants and factories were to be financed or refurbished, involving an expected $380 million. Some 50 000 workers were to be employed. The time frame for the plan was the year 2000. This plan was certainly over-optimistic in its scope.

26. As reported in *Jingji Daobao*, Hong Kong (in Chinese), 20th June 1983.

27. These include: a plastic toy factory, a clothing factory, a spare parts production unit, assembly operations in the areas of leather goods, cardboard boxes, aluminium products, a cigarette factory, plastic goods, hand made garments. See *Jingji Daobao, op. cit.*

28. For a full review of the Shantou incentive package, see *SWB*, FE/W1231/A/10, 13th April 1983.

29. "Recent Developments in Shantou Special Economic Zone", Huang Biaoxing, *China Market*, No. 10, 1982, p. 75.

30. As reported in *SWB* FE/W1231/A/11, 13th April 1983.

31. Premier Zhao Ziyang declared in Xiamen in September 1983 that SEZs might have to consider providing new incentives in order to attract foreign investment, including the opening of the domestic market. *Wen Wei Po*, 30th September 1983.

32. *Investment Guide to Shantou* SEZ, Shantou, May 1983 (in Chinese).

33. See for example *Shenzhen Special Economic Zones: China's Experiment in Modernization*, Kwan Yiuwong, (Ed.), Hong Kong, 1982; *The Largest Special Economic Zone of China – Shenzhen*, David Chu (Ed.), Hong Kong, 1983 (in Chinese). Numerous articles are also devoted to Shenzhen in the Chinese, Hong Kong and European press. Most of them simply describe the economic development of the zone. Critical articles have begun to appear in the local Hong Kong press as well.

34. See *Guide to China's Foreign Economic Relations and Trade: Investment Special*, Hong Kong, 1983. The publication gives the text to all laws relating to foreign economic relations (in Chinese and English).

35. The new status of a "shi" (municipality) granted to Shenzhen should not be confused with the same term (in Chinese pinyin transcription) used for a municipality that has the administrative status of a province. Beijing, Shanghai and Tianjin enjoy this degree of independence.

36. In this regard, see the recent study *China and Hong Kong: The Economic Nexus*, A. Youngson, (Ed.), Hong Kong, 1983. The chapter entitled "On Some Problems of Modelling Chinese Economic Growth" by Fu-Sen Chen and Tow Cheung contains a table (pp. 202-203) on the industrial wages and cost of living index from Hong Kong, 1953-1980.

37.

Liantang	2.0 sq. km.	Shuiba	0.4 sq. km.
Bagualing	1.06 sq. km.	Shangbu	1.2 sq. km.
Shale	1 sq. km.	Futian	3 sq. km.
Chegoingmiao	2 sq. km.	Dashahe	2 sq. km.
Shekou	1.2 sq. km.	Nantou	1 sq. km.

38. See Chu, *op. cit.* It was reported that from January to June 1984, over Rmb 1 000 million had been spent on infrastructure by the Chinese authorities. Cumulative basic construction investment from 1979 to June 1984 was reported to be Rmb 2.5 billion. Cf. also Ho and Huenemann, *China's Open Door Policy: the Quest for Foreign Technology and Capital*, University of British Columbia, Vancouver, 1984, p. 71 for a reconstructed estimate by zone.

39. Cost estimated HK$ 2.4 billion (joint venture with Hopewell Holdings Hong Kong). It was scheduled for completion in 1985, but will certainly take several more years.

40. For a complete description of the Shekou area, see the *Investor's Handbook* (Revised Edition), Shenzhen, China Merchants Shekou Industrial Zone, Shenzhen Special Economic Zone of Guangdong Province of PRC, 1983, pp. 50-53.

41. The CMSN was created by the Qing Government in the 1870s as a state-sponsored shipping company to take over the private foreign firms that had until then dominated the Yangtze and China Coast routes. The firm's chief competitors were the Hong Kong based Jardine-Matheson and Butterfield and Swire. See Hou Chi-ming, *Foreign Investment and Economic Development in China, 1840-1937*, Cambridge, Mass., Harvard University Press 1965, pp. 59 ff.

42. CMSN Company spent more than $100 million on infrastructure in the period 1979-end 1983.

43. As reported in the collective work in *Xueshu yanjiu*, No. 1, 1982 (in Chinese). This is also discussed in the investment guide to the Shekou industrial zone, and was reiterated during interviews with officials of the CMSN during visits to the Shekou zone in December 1983.

44. "The Take-off of the Industrial Zone of Shekou" (collective authorship by the research staff of the CMSN) in *Xueshu yanjiu* No. 1, 1982 (in Chinese).

45. See *Xueshu yanjiu*, No. 1, 1982, as well as the presentations made in the *Investment Guide to Shekou: Shekou Industrial Zone of China's Merchant Steam Navigation Co. Ltd. in Shenzhen Special Economic Zone*, September 1983, Shenzhen, PRC, p. 50.

46. As reported in the Hong Kong press; "Local Investment in Shekou zone over $100 million", Olivia Sin, *South China Morning Post*, 28th August 1983.

47. Among the advantages listed were:

   a) freedom from "interference" vis-à-vis the Chinese administration;
   b) a means of better understanding the modern methods of economic management, including a better understanding of the "profit" mechanism;
   c) a means of exploring the possibilities of reforming the Chinese system of salaries, housing allotment and new incentive systems;
   d) making the principle of "self-sufficiency" better understood and applied within the Chinese economic framework.

   See *Xueshu yanjiu, op. cit.*, pp. 12-13.

48. A joint venture with: CMSN, CNOOC, CNOJSC, COOS (HK), China Development Finance (HK), Shenzhen SEZ Development Co., China Resources Holding and Huang Zhen Hui Investment Co. (HK).

49. The cost of the joint venture is estimated to be $4.6 billion with the Hong Kong based China Power and Light Company as major shareholder of Hong Kong Nuclear Investment Corporation.

Chapter 5

THE SPECIAL ECONOMIC ZONES:
PERFORMANCE AND IMPACT
ON THE CHINESE ECONOMY

The SEZs have not had uniform economic development, nor has their performance been equally successful. It may be too early to judge the impact of the zones on the Chinese economy in general; however, it is certainly pertinent to review the first four years of SEZ performance in the light of Chinese expectations, foreign business interests and the international community at large, and relate these to development strategies and options for China. Evaluation of the success or failure of the SEZs depends a good deal upon the point of view of the observer, and for this, it is necessary to examine three different positions: that of the Chinese authorities, that of foreign firms, and that of foreign governments.

In creating the SEZs, the Chinese authorities made it clear that the zones were part of an overall strategy to modernize the country by importing at minimum costs – both ideologically and economically – the technological skills and equipment of the industrialised countries. The zones were designed in part to provide new inputs to domestic industries, and, in a lesser way, provide consumer goods that were currently being imported; the commercial balance was also a concern. Thus, both an import substitution and export promotion strategy were cited frequently by Chinese authorities in justifying the creation of the SEZs.

This double thrust is not without problems. How can the central authorities evaluate the success of the SEZs on both these fronts at the same time? Export strategies require massive production investments in internationally competitive industries. The PRC has few such industries; the majority of the exports of the country in 1983 fell into the category of light manufactured industrial and textile goods, mineral fuels and heavy industrial and chemical products. The SEZs have not been successful in attracting industry in general, and the investment share for light industry for the SEZs has been relatively low in comparison with real estate investment. Chinese authorities will certainly be evaluating the export performance of the SEZs in light of increased export volume for SEZ production. On the other hand, the advocates of an enclave economy see new technology (whether high technology, as was first announced in the 1980 laws or simple appropriate technology as is implied in the second set of laws, promulgated in September 1983) and management skills as another set of priorities; these they said, might be acquired while at the same time a proportion of China's technology imports might be reduced by foreign joint ventures. Government authorities will be studying carefully the kinds of industries that have been attracted to the SEZs, and whether these industries have effectively been able to reduce import dependence in key sectors of the economy. They will also be anxious to know whether foreign investors have brought with them new advanced technologies involved in the production of essential machine equipment or capital goods for the industrial sector; they will also be monitoring consumer goods production for the domestic market. Chinese authorities will be keenly interested to know how much of the management and organisation techniques have been transferred inward to the national economy from the SEZs.

Effective means of measuring SEZ successes are difficult to establish. Some indirect criteria do exist. The number of joint ventures with provincial or municipal Chinese enterprises is an indication of the interest expressed in the SEZ experiment by national groups. The methods for recruiting and allocating workers to the SEZs and their reassignment afterwards are important indicators of the possible transfer of skills learned on the job; structural reforms of the enterprise system that have experimented within the SEZs and imported into the rest of the national economy are also indications of the SEZ success. The hard line accountants will want to measure capital construction and infrastructural costs in the SEZs against revenues, cost-sharing with foreign enterprises, and export earnings. Virtually everybody will be monitoring the export performance of the zones, whether it is towards the domestic economy or towards the international market. Technology acquisition is a further indicator of successful SEZ investment-incentive policies.

Foreign business concerns are primarily interested in lowering their production costs and therefore will be calculating unit costs in the context of alternative sites, production procedures, management techniques, skill levels and wages. Most production firms with foreign capital assets are also anxiously awaiting further information concerning the opening of the domestic market to production both within the SEZs, and from beyond. Businesses will be asking themselves whether it is not worth the trouble to become familiar with Chinese market conditions from within the SEZs in order to be better prepared for the day when the domestic market is opened to international commerce. Many feel that the investment they make in the SEZs is a vote of confidence in the open door policy in China, and that this stake in the success of the experimental zones will win them friends in the Chinese hierarchy who will later be useful when commercial operations become more common within the PRC. These firms will not only be looking at profits and losses; they might see their presence as an investment in the future, ready to write off current account losses for future access gains. The logic of this cost-benefit operation is quite different from that of the smaller firms whose primary evaluation of success will be on the returns on investment. These latter, the majority of them Hong Kong Chinese investors, do not see an expanded role for themselves in the domestic economy, but are rather trying to avoid rising costs elsewhere by relocating within the SEZs. Many of these firms have three to five-year investment horizons and are seeking short- to middle-term gains.

The political cost-benefit analysis, both on the part of the Chinese government and on the part of foreign governments is yet another logic. The Chinese authorities may well have taken considerable risks in promoting the idea of foreign investment zones within China, as recent literature indicates. The reform movement in China seems well advanced, but there is a long procedure of consensus building concerning the role of foreigners in China, and the opening of the country to a foreign economic presence. The administrative and political cadres who owe their position to the Maoist regime are reluctant to see their own positions publically vilified by the proponents of the open door policy. This group, which constitutes a significant minority within the administration and the army, will be evaluating the SEZs as channels of "spiritual pollution" or "deviation" from the socialist goals of the PRC. Their influence in the party is an important factor for reform-minded administrators in Beijing. The thin line between modernizing in a Chinese fashion and adopting a wholesale foreign model must be maintained at all times for the political authorities in Beijing. The absolute need to use the SEZs and other semi-market experiments as channels through which new technologies and new techniques will flow is evident to all who wish to quicken the pace of development in the country; but the risk is equally apparent: too much, too fast, could introduce once again the chaotic conditions (wide income differentials, poverty, high unemployment, a weakened state apparatus, inflation, balance-of-payments deficits) that brought about the rise to power of the Communist Party in 1949.

Foreign governments have different views. Some of the Pacific region industrial countries wish to see China embark upon a stable economic programme that will bring the PRC more into the realm of international economic and political affairs. Their primary concern is to foster a transition between the isolationist position of the Maoist era towards an integrated economic presence of China in the region. The present Chinese leadership is committed to bringing China through the threshold of the open door; some industrial powers are therefore anxious to support the reforms, and have even contributed substantially to them through international loans and semi-official government investment. While remaining prudent about the commercial profits to be gained by installing enterprises in the SEZs, governments such as those of Japan and Australia have shown considerable interest in the experiment, and have participated in numerous activities to promote national investment in the SEZs. While different ministries of foreign governments have different objectives, there is a consensus in the region that the cost of stabilisation is not necessarily to be calculated in terms of commercial profits. These foreign governments are following the SEZ evolution carefully, although it is seen as part of a general evolution of the PRC rather than an isolated experiment.

Other industrial countries have more straightforwardly commercial attitudes about the SEZ experiment. Cost-benefit analysis is carried out in a general economic framework of commercial balances; the SEZs can be evaluated as export processing zones that may dig deeper into the market shares of certain light manufactured goods or textiles. The NICs of the region are also concerned about this potential threat to their own export processing zones and light manufacturing industries.

The policy mix adopted among industrial countries towards the SEZs is a series of weighted doses of concern for commercial balances, future evolution of export commodities and basic political stability.

When the SEZs first opened in 1980, the choices offered foreign firms who wished to establish themselves in China were few. No legal structure of any type protected their interests; and there was no clear idea of how much – or for how long – China was committed to open herself to the outside. The political evolution of the last four years has been dramatic indeed. Not only have the SEZs expanded into fully operational zones, but many other areas of China have opened to the possibility of foreign investment, including the 14 coastal cities opened in May 1984. This situation has changed the role of the SEZs. They no longer are the only possible areas for investment, although they remain the only Chinese territory with special investment laws for foreigners. Now foreign investors have the possibility of choosing a site from among a number of possible cities, some with considerable industrial bases, and long experience in exporting onto the international markets. Certain advantages of the SEZs are being spread to other areas of the country, although no other region has received the full package of incentives that have made the SEZs unique. It may not be long before they do, however.

SEZ PERFORMANCE

By 1985, the four SEZs had all reached the start-up stage; foreign direct investment had been attracted to some degree in all four zones. However, Shenzhen experienced the lion's share of growth, due in great part to the proximity to the Hong Kong economy. Zhuhai attracted largely infrastructural investment, and until the formation of the Everbright Development Company, was moving sluggishly. When Everbright took over the promotion of the industrial parks of Zhuhai, a more orderly plan was developed, and investment began to pick up. Table 39 gives the evolution of foreign direct investment for all four zones through December 1984. The table includes the large $1.5 billion pledged investment contract for Zhukai which appears to have been put aside by 1985. Considerable variations occur in official statistics for the SEZs, but the table may be considered a conservative estimate of aggregate foreign direct investment.

Table 39. DIRECT FOREIGN INVESTMENT IN CHINA'S SEZS (CUMULATIVE)

Unit: $ million

SEZ		Up to December 1981			Up to June 1983			Up to June 1985		
		N	CP	CR	N	CP	CR	N	CP	CR
Shenzhen		1 010	1 463	n.a.	2 266	1 485	286	4 700	2 800	830*
of which:	Shekou	(29)	(88)	n.a.	(70)	n.a.	n.a.	(167)	(200)	(140)
Zhuhai		9	84	n.a.	n.a.	n.a.	29	n.a.	1 680	240
Shantou		n.a.	106	n.a.	n.a.	n.a.	n.a.	n.a.	n.a.	198
Xiamen		5	74	n.a.	n.a.	161[1]	n.a.	153	545	n.a.
Total		1 024	1 727	–	–	–	–	–	–	–

* Figures for Shenzhen are up to end. October, 1985.
N = Number of contracts.
CP = Capital pledged.
CR = Capital engaged.
1. In November 1983.
Source: Compiled from various official sources by OECD Development Centre.

Shenzhen emerges as the most important SEZ, although Zhuhai is not far behind in pledged investment. This is because both zones are *new* investment areas, where several large contracts have signed for infrastructure development. In the case of Zhuhai, this accounts for more than half of pledged investment. Another factor is that both Zhuhai and Shenzhen are in a large measure planned as complementary to the existing external economies of Macao and Hong Kong, and have received numerous contracts for relocation of light industrial production units and recreational facilities. This accounts for the overwhelming initial Hong Kong investment.

The investment performance of the SEZs can be compared to the total of foreign direct investment in the PRC.

Table 40 gives figures for the period January 1979 to December 1984 concerning China's use of foreign capital. It shows that FDI was but a small portion of the new external capital available in the PRC during the period in question; it also shows that FDI share in new external capital has grown proportionately (14 per cent of external capital for the period 1979-82, but 21 per cent for 1979-84). FDI represented, therefore, 47 per cent of *new* external capital introduced into the PRC in 1983, and 54 per cent of the new capital introduced in 1984. The trend, therefore, is towards using FDI as a principal source of external capital to finance modernization.

The breakdown for direct foreign investment is given in Table 41. The figures show that the majority of pledged FDI was in co-production and offshore oil operation (84 per cent of

Table 40.   CHINA'S USE OF FOREIGN CAPITAL: 1979-1984

Unit: $ billion

Items	1979 to 1982[1]		1979 to 1983[2]		Jan. 1979 to Dec. 1984[2]		Jan. 1979 to June 1985[3]	
Loans	10.9	(86 %)	11.9	(82 %)	13.1	(77 %)	13.9	(74 %)
Direct foreign investment	1.7	(14 %)	2.6	(18 %)	4.1	(23 %)	4.8	(26 %)
Total	12.6	(100 %)	14.5	(100 %)	15.8	(100 %)	18.7	(100 %)

1. *JETRO China Newsletter*, No. 51.
2. MOFERT; *SWB*, April 1965; *1984 Almanach of China's Foreign Economic Relations and Trade*.
3. *China Economic News*, No. 29, 1985.
Source: Intertrade, October 1983.

Table 41.   CHINA: BREAKDOWN OF FOREIGN DIRECT INVESTMENT

Unit: $ million

Item	1979 to 1982			1979 to 1983			1979 to 12/1984		
	N	CP	CR	N	CP	CR	N	CP	CR
Joint venture	83	141	103	190	321	173	362	331	n.a.
Co-production	792	2 726	503	1 047	2 950	758	1 372	3 500	n.a.
Offshore oil	12	999	486	23	2 040	788	31	2 400	n.a.
Other[1]	905	1 092	650	998	1 400	966	1 137	800	n.a.
Total	1 792	4 958	1 742	2 258	6 711	2 685	2 902	7 031	3 300

CR = Capital realised.
CP = Capital pledged.
N = Number of contracts.
1. Includes 100 per cent foreign-owned enterprises.
Source: See Table 40.

FDI pledged for 1979-June 1984). Co-production and joint venture is also the predominant form of FDI in the SEZs, as Table 41 indicates.

How much, therefore, of the new external capital in the PRC for the period 1983-84 can be attributed to the SEZs? The answers seem to be surprisingly low. For the period January 1979/December 1984, FDI in the SEZs accounted for little more than 15 per cent of total pledged foreign direct investment for China. This is not surprising, if offshore oil investments are taken into account; these latter involve large initial investments, and initial outlays. Most of the exploration contracts are not in the SEZs. In the overall performance of the PRC, the SEZ therefore represented but a modest impact upon the growing inflow of foreign capital.

FOREIGN DIRECT INVESTMENT IN THE SEZS

Due to partial data for Xiamen, Shantou and Zhuhai, it is difficult to generalise about FDI patterns in all of the SEZs. For reasons mentioned above, it is clear that Shenzhen and Zhuhai are fundamentally different in their vocation as new investment zones; Shantou and

Xiamen, which have large import substitution possibilities, both wish to attract foreign direct investment into existing industries. These are different investment strategies altogether. As Shenzhen is the most developed of the SEZs, and its investment and performance data are the most complete of the SEZs, the following remarks will be limited to the Shenzhen experience, with appropriate remarks when necessary for other SEZs for which relevant data exist.

Table 42 shows some important trends in the SEZ. The first is the dramatic growth of the gross value of industrial and agricultural output from Rmb 1.47 billion in 1978, before the industrialisation of the SEZ, to Rmb 8.7 billion at the end of 1983. The majority of the increase came from industrial output (12 times the output in 1984 compared to 1978). At the same time, these figures must be compared to the investment figures for basic construction. It becomes clear, then, that the basic construction investment is correlated to increases in the gross value of industrial output. In fact, the table suggests that without the large infrastructural investment, industrial output would not have been nearly as large. In the calculation of the GOVI, there are no figures available for the part of GOVI which basic construction represented. If the assumption can be made that basic construction is included in the calculation of the GOVI, it is important to determine the exact amount represented by industrial production. In a recent article, the figure for 1983 industrial production for Shenzhen (minus basic construction) was estimated to be only Rmb 120 million, a low level indeed[1].

Table 42. MACROECONOMIC PERFORMANCE FOR SHENZHEN MUNICIPALITY
OVER THE PERIOD 1978-JUNE 1984

Basic Statistics of Shenzhen County

	1978	1979	1980	1981	1982	1983	1984
GOVIA	1.47	1.47	1.90	3.5	n.a.	8.7	n.a.
GOVI	0.6	0.606	0.844	2.43	3.62	7.20	5.55
GOVA	0.87	0.86	1.04	n.a.	1.41	1.5	n.a.
Basic construction	n.a.	0.49	1.24	2.70	6.33	8.86	6.03
Realised FDI (HK$ 100 million)	n.a.	1.2	2.5	5.9	8.8	11.3	5.9
Revenue (govt.)	0.2	0.35	0.55	1.2	1.63	2.96	1.9
Foreign exchange earnings	n.a.	0.28	0.47	0.40	0.56	0.671	1.05
Retail sales	1.17	1.48	2.2	3.5	5.54	12.51	8.61
Exports ($100 million)	n.a.	n.a.	110	n.a.	n.a.	210	n.a.

Unit: Unless otherwise stated, in current Rmb 100 million; for foreign exchange earnings, it is not known what exchange rate was used. Figures are not cumulative.

Source: Adapted from Chen Wenhong, op. cit.
South China Daily News, 14th August 1982.
Economic News, 27th June 1984.
Shenzhen Te Qu Ribao, 8th July 1983; 8th April 1984; 19th May 1984; 8th July 1984; 20th August 1984; 19th May 1984; 14th October 1984.
Yueshu Janjiu, No. 4, 1982; No. 5, 1984.
Maoyi Gouji, No. 5, 1984.
SEZ Nianjian 1983, 1984.
Gang-Ao Jinji, No. 1, 1982.

If it can be stated that Shenzhen industrial output has been modest in absolute terms, and that it was made possible only through massive efforts at capital formation, it can also be argued that this costly first step towards industrialisation cannot yet be judged in terms of output. The stage, rather, has been set in Shenzhen, and the real test of success will be in the years to come, when basic construction is finished and factories are completely operational. In this sense, Shenzhen reached industrial take-off only in 1982-1983; it was also, according to figures, the period in which the GOVI surpassed the GOVA, thus transforming the Shenzhen economy from an agricultural base to an industrial base.

In terms of foreign direct investment, there are two pertinent remarks: first, the ratio of realised FDI to pledged FDI in the SEZs is rising rapidly. In 1983, the ratio of realised capital investment to pledged capital investment was 17 per cent; in 1984, that ratio had risen to 36.7 per cent for all of the SEZs. However, caution must be used with these ratios as this sharp increase may be caused by the extension of the boundaries of three of the SEZs, rather than by more concentration of capital inside the original zones themselves. Secondly, figures show that Chinese basic construction costs represent twice the amount of capital as foreign direct investment in the SEZ. This may be considered normal, however, as the initial start-up costs for the SEZs were considerable, due to the deserted site, and the lack of any infrastructure at the time of its creation. Ten central ministries and central government committees invested more than Rmb 600 million during the period January 1980/September 1984, making possible the quick pace of construction in the zone. These ministries were not only following the advice of the government to invest in the SEZs; they were making sound investments in an area where they could later hope for good returns on investment, and outposts and experimental stations for their own industrial and managerial reforms.

Overall, foreign direct investment accounted for less than 30 per cent of the total investment in the SEZ.

JOINT VENTURES AND OUTPUT

Firms established with foreign capital inside of Shenzhen SEZ (including joint ventures, 100 per cent foreign-owned firms, co-production and assembly operations) accounted for more than 50 per cent of the industrial output of the SEZs in 1984. Here, however, the trend of output is important to note. According to certain estimations[2], the output value of domestic Chinese joint venture enterprises located in the SEZs is rising rapidly, registering greater progress than enterprises with mixed Chinese – foreign investment or 100 per cent foreign-owned. So whereas the share of industrial output may be larger for Sino-foreign joint ventures, the trend is towards increasing the output share of the domestic Chinese enterprises. This could mean that in the future, domestic enterprises will surpass foreign joint ventures in output value mitigating arguments that the foreign joint ventures are essential motors to the SEZ economies.

With regard to the above-mentioned trend, it should be noted that double counting procedures may exist for output value. Domestic Chinese enterprises and joint ventures, are generally tabulated separately by the Chinese officials. In fact, domestic Chinese enterprises may enter into joint ventures with foreign firms as well, and their output value be tabulated under both headings.

Table 43 gives Shenzhen figures for 1984. It is of interest to note that the percentage increase of output value for joint ventures was still higher than that of domestic Chinese enterprises, although once again, it is not certain that double counting has not taken place for Chinese enterprises. From Table 42 it is also clear that retail sales account for Rmb 1 251 million in 1983, or more than 50 per cent of the GOVIA. This would suggest that sales and commercial activities were more important than industrial or agricultural output, something that mitigates the Chinese authorities' claim that the SEZ is primarily an industrial zone.

Table 43. SHENZHEN'S OVERALL STATISTICS IN NOVEMBER 1984

Unit: Rmb million

Item	January/ November output	November output	Change (%) over January/ November 1983
Total industrial output	1 362.25	193.69	122.8
Of which: Light industry	1 121.19	164.83	131.3
Heavy industry	241.06	28.86	90.4
Industries under the arrangement of wholly-owned foreign ventures, Sino-foreign joint ventures and Sino-foreign co-management ventures	674.66	99.69	194.0
Industries within the Special Economic Zone	1 077.2	164.43	120.4
Processing and assembling with supplied materials	114.03	11.27	24.4
Enterprises under the arrangement of domestic co-operation	258.77	35.92	141.1

Source: State Statistical Bureau, 1984.

There is also a positive correlation between the increases in GOVI and that of retail sales in general, ending in 1983.

Table 42 also gives an important indication of export value for the SEZ. Shenzhen has only doubled its export capacity in three years, and the total value remains modest; it represented about 1.3 per cent of total PRC exports in 1983. This is in great part due to the fact that Phase I of SEZ development, which ended in 1982, did not really see many operational enterprises in the SEZ. However, should the zone be able to continue to develop export capacity at the same pace over the next ten years, it could have a significant impact on overall export performance.

Transit trade plays an important role in the overall economic life of the SEZ. Given the data from the commercial and financial purchasing systems of the SEZ, figures show that 24.4 per cent of merchandise bought by the zone was made in the SEZ; 25 per cent was acquired by exchange agreements, or co-operative agreements (such as the firms in the SEZ purchasing with foreign currency needed technology and then exchanging it with inland provinces for agricultural goods); 50.6 per cent was imported from abroad, representing Rmb 600 million in retail sales.

In the same year, the SEZ purchased about 50 per cent of its production capital goods from abroad, representing Rmb 360 million. This is in the same proportion as the commercial system purchases mentioned above.

It has been estimated that about Rmb 600 million worth of goods travel through the SEZ to the domestic market per year[3].

In 1983, market estimations for Shenzhen showed surprising patterns of behaviour within the SEZ. It was estimated that 40 per cent of buyers on the Shenzhen market came from the SEZ; 50 per cent came from other regions in China, and only 10 per cent came from outside China. In some product lines, such as electronic goods, it was estimated that more than 80 per cent of the sales were to Chinese from within the PRC. Given the per capita retail sales of Rmb 4 170, this makes Shenzhen one of the wealthiest regions of China when compared to Shanghai (Rmb 912) and Beijing (Rmb 896) or Guangzhou (Rmb 504). This can be explained by the large number of "external" Chinese buyers on the SEZ market, and therefore a strong backward link to the domestic Chinese economy.

Transit trade was bound to develop in the SEZ, given the way in which the zone was set up. As raw materials and component parts for production enter into the SEZ tax free, and are either produced or assembled in Shenzhen, they have effectively entered in the PRC tax free, as little or no control exists between the SEZ and the rest of China. Industrial products either originating or passing through Shenzhen are still relatively cheaper than equivalent quality goods made within the PRC, and thus transit is implicitly encouraged. This situation will surely change once the border fence between Shenzhen and Guangdong province is utilised.

There is also an administrative factor that favours transit trade. The traditional planning bureaux within the PRC have limited transit trade within the country; as these latter do not exist in the SEZ, the market allocation mechanisms work in favour of a transit trade with the rest of China. This practice has led to the appearance of illegal trade and blackmarket for currency exchange and consumer goods.

INVESTMENT INCENTIVES IN THE SEZS

The incentive package devised for the SEZs was first announced for Shenzhen, Zhuhai and Shantou SEZs in 1979. The principal features are outlined in Annex 3. The fundamental incentive system was designed to attract foreign investment with the following elements: streamlined administrative control, relative independence for local planning authorities, direct access to provincial and central level planning units, to holidays, duty free allowances on production materials, flexibility in hiring and firing workers, depreciation allowances, negotiated limited access to the domestic Chinese market for goods produced within the SEZ, residence and work permits for foreigners and income tax exemptions for foreigners working within the SEZ. The incentive package was a radical departure from the then current practices in China. This incentive package has evolved over the period 1979-1985, for the most part by lowering taxes and extending duty free status to new categories of articles imported into the SEZs. By 1984, Chinese authorities had made it clear that the incentive packages that had been established for the SEZs was a minimum guideline, and that additional changes could be negotiated in the context of each agreement. Although full details are not available of existing foreign firms' contractual arrangements, it has been reported that several joint ventures have been able to negotiate more favourable conditions because of the promise of new technology transfers, or larger amounts of investment.

Surveys taken among Hong Kong investors in the SEZs and elsewhere indicate that the incentive system in the SEZ is a strong attraction for investors who are considering small- to medium-sized investments with returns on investment of three to five years[4]. This information correlates well with data concerning the average size of Hong Kong contracts within the SEZs, and the types of investment undertaken, where that information is available.

Recent studies have also demonstrated that the single most important incentive for foreign investors in developing countries is political stability, overriding even lower wage, manpower, land utility and fiscal considerations. The SEZs therefore were among the first areas of China to introduce special legislation to ensure the image of a stable, more neutral political environment[5]. By 1981, the administrative cadres of the Shenzhen SEZ were clearly being chosen for economic and managerial experience; political activism was played down and, indeed, later discouraged. This does not mean, however, that administrative cadres do not have to be inspired by the socialist economic theory that China is seeking to develop within the SEZs. The new "mixed capitalist and socialist" system set up in the SEZs also requires cadres to remain in line with directives coming from central authorities. The political and ideological elements of daily life are simply played down; they have hardly disappeared.

The reform of the managerial system for state enterprises disclosed in 1984 makes use of experiments in separating administrative and political functions that were pioneered in the SEZs. These measures can be seen as efforts to recuperate younger managers, left out of the political system during the Cultural Revolution and at the same time create a permanent administrative structure which will be separate and parallel to the party recruitment and advancement system. This move should be seen in the context of the Deng team's efforts to overcome resistance within the party, and to allow younger technical professionals to move into positions of authority and responsibility without advancing at the same time within the party apparatus.

INVESTMENT PATTERNS

Figures for Shenzhen SEZ show that the majority of pledged foreign direct investment in the first phase of the SEZ development (up until the end of 1983) was in the area of real estate development, principally hotels, apartment units and commercial space to be rented.

Real estate investment often was negotiated in phased packages over a number of years; pledged figures are aggregates of all phases, and therefore represent, in fact, multiple investment agreements projected over several years; industrial investment, on the other hand, requires immediate capital investment, representing the majority of the operational costs to set up production units. Investment figures, therefore, often represent shorter periods of time, and a greater proportion of the pledged investment must be realised in order to begin reaping returns on investment. The large-scale real estate investment is partially explained by the 1980-1981 boom in Hong Kong real estate markets, as well as the fact that Hong Kong investors were interested in procuring housing for extended family members who live in China and who wanted to relocate in Shenzhen. The fact that hotel investment reaped foreign exchange profits which could be repatriated without passing through foreign exchange operations with the Bank of China also made such investment attractive to foreign firms and individual investors.

Investment figures for the year 1985 only were released by the Shenzhen municipal authorities in early 1985. They indicate that overall foreign direct investment in Shenzhen was HK$ 5.1 billion, representing 885 contracts; 80 per cent of the contracts were reported to be for industrial projects, a sizeable increase in the proportion of investment for the zone. However these figures must be approached with caution. In giving figures for FDI and the number of contracts, without giving the investment figures per contract, it is not sure that the same proportion of 80 per cent would pertain for total investment capital pledged for industrial projects. It is also unsure whether the number of 885 contracts represents all new contracts, or new and renegotiated contracts, something that would amount to double counting for contracts that were negotiated earlier in the period 1980-1983.

The ratio of realised investment to pledged investment has been on the rise in Shenzhen as well, perhaps due in part to the disbursement of second phase investment capital. This supports the hypothesis that industrial investment, which has higher initial capital investment is in fact rising as well.

The period 1979-1983 should be viewed as a first phase of investment in the SEZs. Investors were prudent in their behaviour; small contracts were concentrated in small- to medium-sized ventures, principally in the area of compensation trade and co-production agreements when they were industrial projects. From 1982 to 1983, industrial investment decreased; by 1984, it had once again registered a sharp rise.

The origin of equity joint venture agreements is an important indication of the geographic distribution of interest in the SEZs. Table 44 presents the origin of equity joint ventures by country. Hong Kong and Macao investors signed the largest number of equity joint venture contracts, although the average size of the contract was much lower than Japanese or American investment. The table also indicates that in the first stage of SEZ investment, Hong Kong and Macao played a significant role in the zones, but that effort probably represented transfer of existing Hong Kong facilities or small contractual agreements to process and assemble. This is also true for PRC investment in the SEZs. The SEZs needed much more the OECD area investors, who signed few contracts, but brought new industries and large contracts to the SEZs. Although attracting Hong Kong foreign direct investment has been part of the political programme of the SEZs, it is also important, for the technology transfers and associated management skills, to attract large OECD area firms with medium to advanced technologies and existing marketing capacities in the international system.

Table 44.   COUNTRY OF ORIGIN OF EQUITY JOINT VENTURES WITHIN THE SEZ, TO NOVEMBER 1984

Country	Number of ventures	Total investment	Foreign investment	Foreign share	Average size	Country share in total inv.
		$ mill.	$ mill.	%	$ .000	%
(1)	(2)	(3)	(4)	(5)	(6)	(7)
Japan	9 (5)*	70.6	42.3	60.0	7 758	26.0
United States	8 (5)*	58.9	29.2	49.6	5 840	18.0
Hong Kong + Macao	33 (32)*	67.7	36.1	53.3	1 128	22.2
Fed. Rep. Germany	1 (1)*	27.5	13.7	50.0	13 700	8.4
Singapore	3 (1)*	20.0	6.0	30.0	6 000	3.7
United Kingdom	2 (2)*	7.7	3.8	49.4	1 900	2.3
Denmark	1 (1)*	5.0	2.5	50.0	2 500	1.5
Philippines	1 (1)*	4.5	2.3	51.0	2 300	1.4
Switzerland	1 (1)	0.36	0.18	50.0	179	0.1
France	2 (0)*	n.a.	n.a.	n.a.	n.a.	n.a.
Hong Kong/1 foreigner partner	3 (2)*	50.9	25.1	49.3	12 500	15.4
Unspecified	4 (2)*	1.5	1.4	91.0	650	0.9
Total	68 (53)	314.66	162.58	51.5	3 008	100

* Number of joint ventures for which figures on foreign investment exist; only those ventures are taken into account for the calculation of columns (3) to (7).

Source: OECD Development Centre data base on Equity Joint Ventures in the People's Republic of China.

In the present study, an effort has been made to isolate equity joint venture figures when dealing with the SEZs in order to determine the impact of OECD area firms on this special type of investment. Equity joint venture is considered apart, principally because it:

1.   involves equity shares that normally involve real capital transfers;
2.   it involves sharing risks and profits on an equity basis, thus associating Chinese and foreign firms on equal footing;

141

3. equity joint ventures often involve management and technology transfers associated with the agreements;
4. it has the most detailed legal protection;
5. Equity joint ventures have the most reliable figures for FDI.

There is also special interest for foreign firms in making use of the equity joint venture formula when investing in China. The domestic market is opened under certain conditions to the products of joint ventures in the SEZs and the rest of China; the PRC itself may encourage the creation of export bases in which foreign joint ventures could play a lead role in exporting goods produced in the PRC; China has a large natural resource reserve, and this stable supply base might serve as an attraction to foreign firms anxious to secure raw materials for production purposes. Lastly, technology transfer can take place under both contract and sale conditions, and joint ventures that are located with the PRC may well have advantage in selling new technologies to the Chinese when these imports are deemed necessary.

The ratio of equity joint ventures to other forms of FDI in China, however, is not high, as Table 41 demonstrates. It is clear that contractual joint ventures in which the share of profits are determined by contract rather than by equity shares are the major form of FDI used by foreign partners in the PRC. Similar figures exist for the period 1981-1982 in Shenzhen and Zhuhai, as presented in Table 45. In this table, equity joint ventures and contractual joint ventures are lumped together. In the SEZs, investment was concentrated in these two forms of agreement, as the table shows.

Table 45. INVESTMENT BY LEGAL CATEGORY, 1982 %

Type	Shenzhen[1]		Shekou[2]	Zhuhai[2]	
	Amount of investments	No. of contracts	No. of contracts	Amount of investments	No. of contracts
EJV and Co-production	82	11.1	88.5	87.2	17.6
100 % subsidiary	15	2.3	11.5	0	0
Offshore oil	n.a.	n.a.	n.a.	n.a.	n.a.
Compensation	n.a.	n.a.	0	8.1	8.6
Subcontracting (processing assembly)	2	86.0	0	3.7	61.5
Other	0.1	0.6	0	0.8	12.3

1. In 1981, from Steffani, "Prospettiva della Planificazione economica Pragmatica in Cina", 1983 *Mondo Cinese*.
2. Official sources interviewed by the author in SEZ's, 1983.
Sources: Compiled by OECD Development Centre from various sources.

Moreover, for China in general, there is a growing tendency by foreign partners to enter into forms of FDI involving more risk sharing by the latter, including equity joint ventures, and 100 per cent foreign-owned firms, as opposed to compensation trade, buy-back agreements, processing and assembling.

Table 46 gives a breakdown of firms in Shenzhen by sector. It is clear that manufacturing enterprises are in the overwhelming majority. Among these, co-production, assembly and processing represent the largest number of contracts. This can be deducted from the comparison of figures for equity joint ventures and those for other forms of FDI in the SEZs.

Only partial data exist for sectoral investment, and it is too early to speak of definite sectoral patterns of FDI in the SEZs. Initial results for equity joint ventures, presented in Table 47 show that the manufacturing sector has also attracted the largest number of equity

Table 46. NUMBER OF FIRMS IN SHENZHEN SEZ, 1979-1983

Type	1979	1980	1981	1982	1983	Total
Manufacturing	112	242	321	457	741	1 873
Retail trade/restaurant	5	10	5	25	87	132
Transport	3	4	5	1	15	28
Real estate	2	9	25	6	17	59
Tourism	2	5	3	2	4	16
Agriculture, stock farm	46	32	218	86	41	423

Source: Liang Xiang, "Zhenzheng ri shang de Shenzhen jingi tequ", *Renmin Ribao,* 29th March 1984.

Table 47. EQUITY JOINT VENTURE IN PRC
BY LOCATION/ACTIVITIES EFFECTIVE 1979-NOVEMBER 1984

Location Activities	SEZ			Outside SEZ					
	Shenzhen	Zhuhai	Xiamen	Beijing	Shangai	Tianjin	Other coastal cities and Hainan	Rest of the country and unknown	Sub-total
Primary	4				1			3	8
Agriculture									
Fishery and animal husbandry	1				1			3	5
Mining and quarrying	3								3
Manufacturing	41		6	12	6	16	14	41	136
Machinery									
Electronics and electrical appliances	7		2	5		2	2	9	27
Automation					1			3	4
Precision	1					1	1		6
General (escalator etc.)	2			2		1		1	6
Food, beverages, tobacco	1		2	1	1	1	3	3	12
Textiles			1			2	1	8	12
Industrial materials	3				1			5	9
Woods and furniture	5							3	8
Construction materials	5							1	6
Film processing		1					3	2	6
Plastics, synthetic fibre	2					1		1	4
Cosmetics					1	1		1	3
Medicine, drugs					1	1		1	3
Metallurgy	2								
Oil survey, service station, etc.	7			3	1	5	2	2	20
Transport equipment	6			1		1	2		10
Service	12	1	2	11	4	4	8	9	51
Real estate	2		1	5	2	1	5	7	23
Catering and tourism	6			3		3		1	13
Sales, trade	3	1		1	1		2	1	9
Leasing (equipment)				2			1		3
Communications	1		1		1				3
Miscellaneous	2			1		2	1	6	12
Total	59	1	8	24	11	22	23	59	207

Note: This partial list gives details for equity joint ventures for which figures exist: Chinese authorities announced in late 1984 that over 360 joint venture agreement had been signed for China. Full details were not available for all joint ventures. By mid-1985, the same authorities indicated – without giving any details – that 741 joint venture (equity and contractual) agreements had been signed for 1984 alone; this brings the total joint ventures signed for China to 970. No breakdown is available.

Source: Compiled by OECD from "Delegates Survey Report on the Investment Environment in China", October 1984, China-Japan Trade Association.

143

joint ventures in the SEZs, particularly in the area of electronic components and transport equipment.

Workforce

One of the chronic problems in the SEZs is the maintenance of an adequate supply of skilled labour. The Shenzhen municipal government set up a special company, the Labour Services Company, which provides foreign investors with candidates. In part, recruitment is done from the local population, which is not skilled labour. In this sense, the population growth and movement in the last five years in the Bao'an County, in which Shenzhen is located, is significant. When the SEZ was set up, it was estimated that the total population of the whole county was 320 000, centred in the city of Shenzhen[6]. The SEZ has a population of 120 000, showing a rapid growth – or more correctly, a relocation – of the county population. In 1978, before the creation of the zone, the area now comprising the SEZ inside the county had a population of 68 166; in 1980, it had risen to 84 057; by 1984 it was more than 100 000 for the same area. This has put considerable strain on the existing facilities within the SEZ, and much of the housing and infrastructure investment has had to go to support the new working population. It is projected that by the year 2000, the total population of the SEZ will have risen to more than 400 000. It is useful to ask where this population is coming from. Much of the unskilled labour force comes from the rural population of the Bao'an county in back of Shenzhen. This rural exodus is part of the effect of higher wages and better living conditions inside the SEZ, even when compared to other urban areas of the PRC. It has caused a shortage of rural labour, and farmers have had to be imported to Bao'an county from the PRC to supply the fresh food crops that the growing population within the SEZ consumes daily. These latter have come from the Chaozhou-Shantou area (a reported 2 000 vegetable farmers imported in 1980), and from the municipality of Guangzhou (2 500 more in 1981)[8]. This movement of labour is facilitated by the organisation of labour recruitment within the SEZ and a relay, through the Labour Services company, with the labour supply mechanisms in the PRC. However, these latter did not readily adapt to the new situation offered by the SEZs, which requires mobile, skilled workers. In the traditional Chinese system, workers often stay within the same enterprise all their working life[9]. The planning of manpower allocation is an intricate part of the central economic planning mechanism, which co-operates closely with the local planning commissions. This information is in turn relayed to educational institutions for planning purposes. After graduation from a training institution, skilled workers are assigned to an enterprise. Should an enterprise need new skilled labour that is not available from the graduating class of workers that year, it applies to the provincial industrial bureau on which it depends. If the enterprise is not able to find qualified workers within industries of the same branch, it applies to the provincial labour board which carries on a search province-wide. Should that fail, the provincial board applies to the Central Labour Board in Beijing, which conducts a nationwide search. This can be a long and involved process, leaving many jobs unfilled and creating bottlenecks in other areas where there is too great a supply of skilled workers.

This situation does not prevail in the SEZs. In Shenzhen, the municipal company which takes on the responsibility for recruitment of personnel also takes on the responsibility of locating appropriate candidates within the national PRC system. This is an important advantage, as it is apparent in conversations with SEZ authorities that recruitment within the PRC has been a vital part of the rapid growth within the SEZs. The labour service companies in all of the zones are able to contract with provincial enterprise boards for the transfer of labour into the zones for periods of time of up to ten years. In some cases, whole factories, including machines and labour force have been transferred into Shenzhen). The People's Liberation Army has also been a significant, if largely ignored partner in the recruitment of labour for the SEZs.

The labour problem can hardly be underestimated, as it remains one of the most serious problems in the Chinese economy, and one of the stated reasons for opening up the SEZs to new forms of international economic co-operation. The on-the-job training that Chinese workers will receive in the enterprises located inside the SEZs is a practical means of creating new forms of education and skill upgrading at low costs. It is not, however, perceived with equanimity by both Chinese and foreign investors.

The situation in Shenzhen is an interesting commentary on the general Chinese situation. The Shenzhen Labour Services Company, in the period 1979-1981 transferred more than 1 200 cadres into the SEZ for management posts. They were selected and brought from other parts of China; but it was reported that only 120 of these had any formal management training[10]. In 1982 alone, between 20 000 and 30 000 "temporary" workers were transferred into the SEZ[11]. In the construction industry alone, the number of workers now exceeds 100 000, all of whom are "temporary" workers, and many of whom are army personnel. Over the last two years, 6 000 technical workers were brought into the SEZ, 10 per cent of whom had qualifications that were higher than an assistant engineer[12].

Another means of bringing workers into the SEZ is for an enterprise within the PRC to contract a joint venture with the Shenzhen Development Company. More than 600 agreements have been made with provinces, cities, autonomous regions and departments of the central government to set up joint ventures. The principal use of these joint ventures is for domestic Chinese enterprises to gain technical experience and a practical knowledge of the workings of the international economy. As these enterprises are in close contact with the foreign partners within the zone, they are able to observe at a close range quality standards, management techniques, and effects of the business cycle[13]. It has been reported that through March 1983, Rmb 649 million have been pledged for joint SEZ-PRC efforts in Shenzhen[14]. This is an important factor in the relay effect of the new techniques and methods learned from the foreign contact within the PRC.

Related to this problem of manpower is the issue of local vocational and educational facilities. The municipal government of Shenzhen founded a university in the SEZ, and the first classes have already been organised[15]. There have been pessimistic forecasts about the role the SEZ enterprises will have on school attendance and teacher employment[16].

Workers are funnelled into their employment through the Labour Services Company; employers have the right to subject workers to tests and trial periods, and fire them if they are found to be unfit. The Labour Service Company takes the responsibility of redeploying workers should they be laid off. There is also a provision in the legal statutes that allows employers to lay off workers if there is a slack in business, or if production must be reduced for economic reasons. The workers are housed as part of their benefits, and social services such as medical care are made available to them through monthly contributions.

Wages

The wage structure in China is complex and is continually evolving. In general, in the PRC, wage scales are set by central authorities, and they differ only slightly nationwide. This system was created in 1956 to ensure even revenue distribution throughout the country. By 1984, the system was being reviewed and substantially modified. The worker scale of wages has eight grades, with a one to three ratio; for technicians and engineers, there is a 16 point scale, for government officials a 26 grade scale, with the widest range of 15 to 1. In principle, the individual worker went up the scale in function of his/her performance. In practice, most stayed at the same level during their professional life (in some cases, individuals advanced at least two grades after entry). This system may now be changing significantly in the SEZs.

The situation in Shenzhen is dramatically different. There is no single wage control within the zone, and bonuses are used widely, as elsewhere in China, but in the SEZs they can account for a larger percentage of income. As one of the avowed purposes of the SEZs is to experiment with capitalist forms of economic management, the Chinese have allowed a great deal of leeway in the wage structure. Until September 1984 the Labour Service Company charged the employer a set fee for each worker; this was determined to be the "salary" of the worker. At least 30 per cent of that sum was kept by the enterprise and municipal government (5 per cent to factory as welfare pension, 25 per cent to government for social costs) as a contribution towards the social security, educational and housing costs in the SEZ. The role of bonuses became crucial here. Some foreign enterprises, in order to stimulate better performance on the part of workers, offered bonuses in the form of Foreign Exchange Certificates, a currency that would allow the workers to purchase imported items in the SEZ. This was not officially permitted, although there is some evidence that it has been practiced. This system of wages underwent many modifications.

Other incentive schemes for workers include the upgrading of status for a worker (rural to urban, urban provincial to SEZ etc.) and the possibility of passing that status down to one's children, implying more spacious housing, and access to better education for one's children.

The average annual wage of a Shenzhen industrial worker rose from Rmb 571 in 1979 to Rmb 2 200 in 1984[17]; farmers have experienced even more striking rises in income (Rmb 685 in 1983, up 329 per cent from 1978). Wages for workers in the joint venture enterprises can be even higher, thus creating an income differential of sizeable importance within the SEZ itself (see Table 48). There have been reports that household income in the SEZ is the highest in China, with a number of households surpassing the Rmb 10 000 a year level[18]. This income distribution problem is linked both to the personal responsibility system that is allowed within the SEZs (as elsewhere in China, but in the SEZs, it is practised intensely) as well as the higher wages brought about by contact with foreign business standards and requirements.

Table 48.   MONTHLY WAGES: SHENZHEN SEZ, 1983

Status	Type/enterprise	Wage to labour company
Farmers	Rmb 57	n.a.
State-owned enterprises	Rmb 113.5	none
Joint venture workers:		
Probationary	HK$ 500	30 %
Contract	HK$ 600	30 %
Skilled	HK$ 700-800	30 %
Technicians	HK$ 1 000-1 200	30 %

Source: D. Chu, "Population Growth and Related Issues in the Development of Shenzhen SEZ", to be published, in *Modernisation in China: Issues of the Special Economic Zones,* Hong Kong, (K. Wong, D. Chu Eds.).

The labour supply is the most often cited problem that foreign investors encounter in setting up enterprises in the zones. The general skill level of labourers is low, and even with the recycling and education courses that are offered to workers, the productivity levels are less than those of Hong Kong. Part of the problem stems from the "iron rice bowl" mentality prevalent in the PRC in which workers are virtually assured employment for life in an enterprise, regardless of performance. Despite recent attempts of the central government to overcome this obstacle, progress is slow. Investors in the SEZs have threatened to close down operations if the promised option of firing inefficient workers is not fully operative.

This rise in wages has, of course, brought about a rise in living costs, despite the centrally planned economy of the SEZ, thus feeding a growing inflation.

In 1981, a wage reform was introduced in the PRC. Replacing the traditional "low wages, low allowances" slogan of the 1950s and 1960s, a new system was instituted in Shenzhen. Three components now made up the worker's wage: the basic wage (jiben gongzi), a "job" wage (zhizu gongzi) and a variable wage (fudong gongzi). The basic wage is the same for any type of employment and is a constant for all workers; the job wage is determined by the nature of the work; and the variable wage is determined by the way in which the job is carried out. This last element is an appreciation of the worker, and could be called a bonus in the traditional system. Recently, comparative figures of this wage system were presented for three different types of workers within the Shenzhen SEZ. The Friendship restaurant in the SEZ breaks down the three types of wages (on the average) in the following manner.

Table 49. WAGE STRUCTURE FRIENDSHIP RESTAURANT, SHENZHEN

Category	Amount (Rmb)	% of total salary
Phase wage	39.22	23.1
Job wage	40.40	24.3
Variable wage	88.38	52.6
Total	168.00	100.0

Source: Louven, op. cit., p. 687, from Renmin Ribao, 11th January 1983.

Table 50. MONTHLY WAGES FOR WORKERS 1982
(ASIA-PACIFIC)

Country	Wage in HK$
Japan[1]	3 950
Hong Kong[1]	1 356
Singapore[1]	1 247
South Korea[1]	1 115
Shenzhen SEZ[2]	540
Shantou SEZ[2]	400
Xiamen SEZ[2]	400
Zhuhai SEZ[2]	400
Guangzhou PRC[3]	200
Philippines[4]	197
China (average)[3]	175

Sources: 1. ILO Yearbook of Labour Statistics, 1983, wages for textile industry.
2. Development Centre estimates based on different official sources.
3. State Statistical Bureau (83), Average Wages of Workers.
4. Far Eastern Economic Review Annual, 1981, Wages for Garment Workers.
5. The official exchange rate was used for each currency.

147

The sizeable portion of the variable wage constitutes an incentive margin for employers. By creating wage differentials within the enterprises, employers can link performance to productivity. The large percentage margin (more than 50 per cent) of variable wage increases control over workers and, as it is not a structural wage, it can be varied each month.

Wages for employees within the SEZ vary considerably. The plant superintendent may receive Rmb 300 a month, a shop floor manager Rmb 200 and a worker Rmb 120 a month. The introduction of the piece wage system has boosted the individual worker's potential considerably[19]. Production workers and administrative workers in Shekou industrial zone are paid on the average higher wages, ranging from a low of Rmb 108 to Rmb 240 a month for production workers to Rmb 181 a month to Rmb 504 a month for administrative workers.

Wages within the SEZs were lower than average wages throughout the Pacific region, in 1982, as Table 50 shows.

Wages were, however, higher than in the Philippines where other export processing zones have been established. Political factors may make the SEZs a better choice for foreign investors than the Philippines.

STATIC AND DYNAMIC GAINS OF THE SEZS FOR CHINA

Two types of analysis must be undertaken when considering the benefits to the PRC of the SEZs. The first is related to balance-of-payments data, and can be measured to some extent with existing SEZ data. The second, and perhaps the more important, is an analysis of the benefits that accrued to the PRC through the intangible demonstration effect, and the transfer of management and quality control skills. In short, the change of mentality the SEZs have worked on the PRC to foster new attitudes towards development goals.

Foreign exchange earnings can be considered one of the measurable gains of the SEZs for China. Foreign exchange receipts are the major means by which the PRC has the capacity to purchase needed imports for domestic production. These inputs are necessary to carry out the ambitious reform programmes that have been announced for the Seventh Five Year Plan. Trade generates the majority of China's foreign exchange. However, balance-of-payments data for the PRC is difficult to locate and analyse, as no complete publication of data exists. Table 51 presents data for 1980.

As indicated in the table, investment capital gains for all of China in 1980 were 4.5 per cent of total foreign exchange earnings. Given the ratio of investment in the SEZs to the rest of China for the same period, it is possible to advance the hypothesis that for 1980 about 2 per cent of gross foreign exchange earnings came from SEZ related investment. Figures for subsequent years will surely alter this when they become available. As SEZ investment has climbed steadily over the period 1980/1984, gross foreign exchange receipts from capital investment may represent more than 5 per cent by 1984. This data, however, represents pledged foreign investment rather than realised investment, and is therefore a maximum figure.

Although SEZ pledged foreign direct investment accounts for more than 50 per cent of FDI in China, FDI itself accounts for too small a portion of foreign exchange receipts to argue that the SEZs are significant with regard to net foreign exchange earnings for the whole country. Should FDI become a major means of attracting foreign exchange, the rising ratio of SEZ foreign direct investment in total investment for China would indicate that the zones are

148

Table 51.   CHINA'S ESTIMATED
GROSS FOREIGN EXCHANGE EARNINGS, 1980

Unit: $ million

Category	Value	Percentage share
Exports	18 130	76.6
Processing and assembly fees	112	
Invisible earnings	2 227	9.4
Tourist revenues and other invisible trade		
earnings	1 200	
Remittances	700	
Export of labour	327	
Capital imports		
Loan receipts	2 260	9.5
Investment capital arising from	1 060	4.5
Compensation trade	100	
Equity/contractual joint ventures	670	
Joint oil exploration	290	

Source: J. Chai "Industrial Cooperation between Hong Kong and China" in *China and Hong Kong: The Economic Nexus*, Hong Kong, 1984.

destined to play a larger role in attracting foreign exchange. At the present time, it cannot be said, therefore, that the SEZs are a significant source of net foreign exchange earnings from investment capital. As little SEZ trade specific data exists, it is too hazardous to try to estimate net foreign exchange earnings from export receipts.

Job creation is another static gain from the creation of the SEZs. Insufficient SEZ specific data exist to carry out an analysis. For Shenzhen, scattered sets of employment figures exist (see Annex 4), but these figures must be viewed with some caution, as they were published at the beginning of the SEZ experiment. By late 1981, it was estimated that 15 000 new jobs had been created in Shenzhen for all types of co-operative agreement; that figure is now estimated to be more than 80 000.

It is reported that the 96 wholly-owned foreign enterprises and the 202 joint ventures located in Shenzhen (November 1984) employed more than 10 000 workers at all levels. These two forms of co-operative agreement tend to have higher job creation per contract than other forms of co-operative agreement, as they involve setting up new production units with new staff, rather than transfer of existing staff to a new production line. It has also been reported that more than 100 000 "temporary" construction workers are in Shenzhen alone, some on active duty with the People's Liberation Army. The problems posed by relocating these workers back inside China after basic construction has finished in Shenzhen have not yet been addressed.

Zhuhai has no substantial industrial employment yet, as the SEZ is still in the initial development stage; Shantou and Xiamen will absorb local unemployment and under-employment with new jobs created in the SEZs. No official figures exist for these latter SEZs, but field work suggested that much of the job creation in Zhuhai, Xiamen and Shantou has been white collar administrative jobs.

To situate the employment gains in an order of magnitude, all types of international co-operation agreements throughout China between 1978 and 1981 were estimated to have

149

created about 598 000 jobs, or only about 2 per cent of the total new urban employment created in the country.

Thus the job creation gains of the SEZs have been small by national standards, although by export processing zone standards, they are large indeed. Gaoxiong in Taiwan, as an example, was estimated to have created more than 50 000 new jobs from 1965 to 1976.

Technology transfer can also be considered to be one of the principal gains of the SEZs. Most technology transfers took place within the context of joint venture agreements, or co-production agreements. Initially, the Chinese authorities aimed at attracting "high technology" transfers, without defining the term carefully. As the incentive package offered to foreign investors depended in part upon technology transfers, it was important to have a precise definition of "high technology". In fact, none of the equity joint ventures in the SEZs required what would be called in foreign terms "high technology". It was necessary, therefore, in revising incentive packages in 1984, to drop the words "high technology" from the text, and grant incentives to firms that merely transferred "technology" which was needed in the SEZ. By December 1984, it was reported that 289 joint ventures and 100 per cent foreign-owned enterprises in Shenzhen SEZ had imported more than 30 000 pieces of modern equipment[20].

Only partial SEZ specific data is available concerning the type, age and origin of technology used in the context of co-operative agreements. There is no reason to believe that there have been large-scale technology transfers into the SEZs as of December 1984; most of the above-mentioned machines are intermediate and low technologies which are considered labour intensive by industrialised countries.

Field research found that much of the technology being used in the context of co-operative agreements was in fact sold to the firm by the foreign partner, or transferred as part of a negotiated equity share. This often meant that the technologies were being phased out of a production unit elsewhere, and being transferred into the SEZ as a labour-intensive technology that took advantage of low labour costs.

Recent research on technology transfers has shown that the success of such transfers depends upon:

a) the degree to which the agreement is an inter-firm, long duration contract for the foreign partner;

b) the degree to which the foreign partner monitors the success of absorbing the new technologies.

On the other hand, Chinese capacity to absorb foreign technology has been estimated to be low. In part, this is due to erratic import policies over the past six years that have left expensive new equipment unused, idling in warehouses waiting for allocation or delivery. This was especially true of the period 1980/1981, when cut-backs in heavy equipment goods were ordered to implement the policy of "readjustment".

It is difficult, however, to measure the absorptive capacity of joint venture firms in the SEZs, as very little data is available. At the firm level, surveys were carried out in Hong Kong in 1981 and 1983 on the basis of Hong Kong investment in Shenzhen. These surveys concluded that Chinese enterprises were slow in training workers for new technologies. The low productivity of workers in the Shenzhen SEZ was attributed in part to the failure to train workers properly for new machines and production tools, as well as the failure to shift quality control standards on the production floor.

Industrial links to the Chinese economy are limited, but seem to be growing. As indicated above, more than 600 joint ventures contracted with PRC state or collective enterprises in Shenzhen alone. In production terms, most inputs used by joint venture firms within the SEZ are imported from abroad; there are also large numbers of assembling and processing operations, where few local inputs are needed. The service sector investment is growing in Shenzhen, and their few inputs are used either from the local economy or from the PRC economy. However, the very large number of co-production contractual joint ventures do use local and domestic Chinese inputs.

In market terms, it has been mentioned above that backward linkages do exist, in the form of transit trade with the PRC. The volume of that trade is difficult to estimate, but important to understand.

PROBLEMS AND ISSUES IN THE SEZS

The first five years of SEZ operations have not generated enough data to speak conclusively about the entire experience. Three of the four zones have not really begun industrial operations; the Shenzhen zone alone provides more than partial information for an informed judgement about the impact of the zones upon the international and the domestic economies. This information, however, must be set in the larger context of Chinese economic reforms, in order to understand the evolving strategies of Chinese leaders in using the SEZs. The evolution is, in some ways, a response to internal criticism of the SEZs.

The first serious criticisms of the SEZs appeared early in 1982, when senior leaders in Beijing openly questioned the wisdom of opening the country to foreign residence and direct foreign investment in "enclaves". These criticisms were not sufficient to call into question the existence of the zones; they did, however, provoke more pronounced official support for the activities. By 1983, other doubts were surfacing: Chinese leaders had looked carefully at the experiment, and before endorsing it, made a number of critical observations. They noted that domestic Chinese investment was much greater than foreign investment, and that the latter was chiefly in the area of real estate. This type of investment, they felt, would not bring in the influx of new technologies that the Chinese had hoped for. They also noted that negative effects were being felt on the neighbouring provincial economies. These latter included uncontrolled immigration into the zone, smuggling and a black market for currencies. There was also a tendency to move too rapidly towards a "capitalist" management structure for the SEZ, in spite of the injunction of the government to take advantage of the opportunity offered by the zones to "get rich". These points were part of the reason why the leadership arguments for the SEZs moved from a purely economic reasoning to a political-economic one.

In November 1983, and again in 1984, senior leaders made official visits to the SEZs. This inspection tour resulted in more adamant official support for the SEZs; and at the same time, it provided the opportunity for the leadership to announce, in April 1984, the opening of the 14 coastal cities of the country to foreign direct investment. It is evident that government officials who support the SEZ experience were able to calm fears that the experiment would spread "spiritual pollution" to other parts of the country as a result of the influence of capitalist ways[21]. This debate is not yet over, in spite of the attempts of the government to play down the negative effects of foreign presence in the country, and the "spread" effects of Shenzhen.

By the beginning of 1985, however, new problems and prospects were emerging for the SEZs. For Shenzhen, the Hong Kong settlement was a boon. Investment began once again to pick up strongly in 1985; figures are not yet available for the types of contracts signed, but according to first reports, industrial contracts have predominated[22]. At the same time, inflation and petty crime have for the first time become a major problem in Shenzhen, and the proximity to the Hong Kong border encourages a good deal of smuggling of consumer goods. SEZ authorities are still trying to master the currency problems[23]. The other three SEZs have not yet had significant foreign presence, and therefore the backlash effects of opening the closed socialist system to market forces have not appeared in the same way as they have in Shenzhen. Zhuhai, due to its location, will certainly experience similar problems. Shantou and Xiamen, whose SEZs are located in middle-sized urban areas, are coastal cities relatively isolated from foreign contact; they may experience more substantial urban immigration problems, once employment is generated by co-operative agreements with foreign firms. They will also have problems of inflation and black marketing of currency, unless the special SEZ money is put into service soon. Perhaps the most serious challenge to the long term success of the SEZs was the decision to open up the 14 coastal cities to foreign direct investment and residence. Once these coastal areas become completely operational, the scarce foreign direct investment will probably move to areas with strong externalities, such as Shanghai. There is also the advantage in cities like Tianjin and Shanghai, of a trained work force and skilled managerial talents. These large coastal areas will certainly be more attractive to large multinational corporations which will need good infrastructure as well as communications with both China and the international market, and an assured supply base for large industrial operations.

To evaluate the impact of the SEZs on Chinese development, it is necessary to look at two levels of issues. The first relate to the classical cost-benefit analysis for the SEZs. Are they economically viable structures within the Chinese economy and are they capable of returns on investment that make them solvent and capable of surviving in the international market environment? Can they indeed be disconnected from the planned economy where resources, including raw materials and intermediate goods inputs, manpower, and capital are allocated by government authorities, rather than market forces? This nascent market economy may be too fragile to bridge the gap effectively between Hong Kong and the rest of China.

A second series of arguments has to do with the more intangible effects of the experiment. The SEZs were established to give China a base for importing new managerial skills, embodied technologies and "learning the capitalist" way. They were to be laboratories where reforms could be tried and once adjusted to the Chinese realities, imported into the country. These bases allowed the Chinese to isolate contact with foreign influence; the SEZs, unlike the 14 coastal cities, are avowedly capitalist economies, where the full strength of market forces is to have its effect. In this sense, the SEZs are still a valuable element in the overall development strategy of the country. While not the mainline of economic development, they allow a space in which unpopular or innovative reforms can be tried, without fear of a spread effect. The long-term political and social cost, then, of the SEZs, is much more important than the short-term balance-of-payments concerns. The SEZs are needed as a window to the world and as a door through which new technologies and skills can pass to China.

Economic Issues

Foreign firms that have expressed interest in investing, or which have actually invested in the SEZs, are vocal in some of their criticisms of the SEZ management and performance. Some of the issues presented below are in the midst of being resolved; others will take significant structural changes in the Chinese SEZ economy before solutions are found.

The principal complaint of foreign firms in the SEZs is related to the problem of determining profits for the joint venture. This key problem has several important ramifications. Theoretically, once enterprise taxes are paid, profits can be converted into foreign currency to be remitted abroad, or used to maintain a foreign currency reserve in the SEZ for the joint venture itself. However, the problem lies at the level of determining the actual level of profits for the enterprise in a given year. Chinese practice is to determine profits by the socialist law of value, a relatively crude measure of the value added. Foreign firms wish to see a much more sophisticated use of Western accounting procedures. However, even when profits have been determined for a joint venture, in practice it has been extremely difficult to remit them abroad. The competent authorities in the Bank of China are slow to approve these transfers, and a great deal of paper work is still required for a routine transfer. Pressure is maintained on the joint venture to maintain the foreign currency reserve in the Bank of China, and recently, added incentives have been announced for firms that keep high foreign currency accounts with the Bank of China[24]. Until the abolition of the internal settlement rate in late 1984, there was little reason to change large sums of foreign currency into renminbi with the Bank of China. Now that the official rate has floated to parity with the internal rate, joint ventures at least do not have to worry about losing money on exchange rates.

The problem of determining profits for the enterprise, however, remains. Certain analyses suggest that profit-making joint ventures are the rule, not the exception, in Shenzhen[25]. According to Chinese sources, in 1983, 148 of the 181 joint ventures, co-operative enterprises and 100 per cent foreign-owned enterprises in Zhenzhen reported to the SEZ authorities at the end of the year on performance; 33 enterprises did not[26]. Table 52 gives the Chinese estimates for profit-making enterprises inside of Shenzhen in 1983; the data is for 103 of the 181 enterprises.

Table 52. PROFITS OF ENTERPRISES IN SHENZHEN IN 1983

Sector	No. of firms		Total sales	Total profits	Average profit rate %
			In renminbi million		
Real estate	8	(8)	36.47	27.56	75.6
Tourism	2	(n.a.)			
Recreation	8	(7)	21.76	4.40	20.2
Transport	2	(n.a.)			
Communications	10	(8)	17.25	2.90	16.8
Industry	35	(27)	72.83	10.27	14.1
Service	42	(n.a.)	.022	.0023	10.5
Total	103	(85)	148.332	45.1323	30.4

Note: Bracketed figures indicate number of profitable enterprises.
Source: Derived from *Jingji Daobao*, No. 37, 1984; adjusted by the OECD Development Centre.

The interest of these figures is the general picture they give. According to these figures, 85 per cent of the enterprises in question were making a profit. The highest profit-making enterprises, not surprisingly, are hotels and real estate, where foreign currency is used for payment by clients. These are also relatively labour-intensive operations for the Chinese. Industries within the zone fared much less well. Their rate of return on investment was 14 per cent; from the statistics presented by the Chinese themselves, it is apparent that "non-productive" investment has the highest returns. As mentioned earlier, hotels, real estate and recreation facilities can request foreign exchange from their customers, thus avoiding the long procedures with the Bank of China for conversion of profits into foreign currency before transferring it abroad. Industries are likewise at a disadvantage in that they also have the difficult problem of determining the real prices for Chinese inputs when calculating the rate of return of investment. This procedure has never been officially explained nor documented.

The profit problem is related to the Chinese reluctance to evaluate projects on the basis of rate of returns. In part this arises from the fear of openly espousing a cost-benefit method that would "put profits in command" or hark back to the "capitalist road". There are few Chinese managers who would risk their careers on promoting complete capitalist management techniques[27]. There are a number of reasons that Chinese managers may not want to use the rate of returns for evaluating projects: the distortion of the price system in determining inputs, the use of large external economies in the form of subsidies for Chinese partners in joint ventures, the lack of experience in calculating rates of return taking into account the above-mentioned items[28].

Foreign firms also complain that profits are eaten into rapidly by the high transaction costs in China. This was confirmed by reports that the United States trade representative had made official requests to the Chinese government to reduce transaction costs in China as a measure of encouraging new business investment[29].

Aside from the issue of profits, there is the effect of the local level costs in the SEZs. These costs are locally added charges or taxes for certain types of services (water, telephone, car rental, etc.) which amount to a "squeeze" by the local authorities. These costs are rather lower for the SEZs than for other areas of China, but they exist in a significant way for Shenzhen.

Unit labour costs, complain foreign firms, are still high, but with the new system of floating wages (up to 60 per cent of salaries), joint ventures hope to raise the productivity of workers within the SEZ. Joint ventures have also noted that a great deal of the risk burden, both in setting up the firm and in marketing the products abroad, has fallen upon the foreign partner, who has had more experience in both management and marketing. As the international market is more competitive than the domestic Chinese market, there have been significant problems in upgrading products for export. With the inability to market domestically, most joint ventures have found that they have to train workers for international standards, thus adding to the costs of the enterprise. This negative factor for the enterprise, is, however, a positive factor for the Chinese, one of whose aims was precisely assimilating quality standards for international export.

Foreign business concerns have adopted a prudent attitude towards the SEZ experience. The larger multinationals have not been attracted to the zones; for the moment, the small- to medium-sized firm is the rule in Shenzhen, principally the result of Hong Kong investment. Many foreign concerns will want to gain China market experience in the now opened 14 coastal cities, and this evolution could dramatically change the fortunes of the SEZs, especially those that have not started operations fully.

Chinese leaders have been less sanguine about the SEZs, and in particular, Shenzhen. Their principal concern is political. The SEZs, even though they may prove costly in an initial stage, are the "Chinese socialist" way of pursuing national development. They remain demonstrations of good will to the international community that China intends to honour commitments to Hong Kong, as well as to foreign business concerns that invest within China itself. They are also convenient ways of monitoring the international economy, through the use of controlled experiments in which Chinese and foreign partners exchange information. The real costs of the SEZs to China may not prove to be the expensive infrastructure that has been installed in Shenzhen and Zhuhai, but the growing social and economic disparities that the new experiment may trigger inside the SEZs. Should those disparities begin to spill over into the national economy before restructuring of the economy can take place, there is every likelihood of a backlash effect against the experiment.

Deng Xiaoping himself has underwritten the essentially political features of the SEZs, deflecting criticism from the cost-benefit economists in the country and the political conservatives. Deng is reported to have seen four essential "window" functions of the SEZs:

1. as a means of absorbing new technologies;
2. as a means of acquiring information about the world beyond, by using foreign skills and techniques, and as a conduit of information about the international economy towards the rest of China;
3. as a means of observing, and absorbing, new management techniques introduced from abroad. This is done through the co-operative agreements signed in the SEZs, as well as foreign training given to Chinese personnel in the SEZ;
4. the SEZs can also serve as a place where special policies, not destined for the rest of the country, are experimented. They can also be used as a means of implementing new foreign policy for the PRC, an "avant garde" of political thinking in the PRC.

In all cases, Deng admonished, "foreign interests must serve national Chinese interests"[30]. These arguments were also used to defend the lead role given to both Guangdong and Fujian provinces in adapting more flexible economic policies, thus relaying many of the reforms of the SEZs to the interior[31].

State Councillor Gu Mu added to the thinking of Deng Xiaoping by providing a more detailed outline of Chinese expectations. Once again, there is a shift away from purely economic criteria to largely political economic criteria. In a long speech to the National People's Congress Standing Committee in January 1985, Gu Mu reiterated support for the SEZs, underscoring four essential reasons for their existence:

1. the SEZs can more effectively solicit foreign capital, introduce foreign technology from abroad, expand foreign trade, and develop the economy by taking advantage of their favourable conditions of being close to Hong Kong and Macao, and by applying the special preferential policy;
2. through the external economic activity of the zones, it is possible to obtain international economic information and to train personnel in various specialised fields;
3. the SEZs can acquire experience for use in China's economic structural reforms. In the SEZs, China can "boldly carry out reform experiments according to the principle of "doing new", undertaking special tasks, and applying brand-new methods while maintaining our stand without change". This, it was remarked, would "enable China to pass successful experience on to the hinterland, and to limit within a small area those proposals which may not prove successful when implemented on a trial basis";
4. "setting up the SEZs has had great political significance.

Public opinion abroad generally believes that is is a creative move. Making the SEZs a success will also produce positive influences for the cause of the complete reunification of the motherland"[32].

The "one country, two systems" use of the SEZs (the changing but not changing, as announced by Gu Mu in number three above) and the political goal of uniting the whole of China, including Taiwan, are long-term expectations of the Chinese. There is every reason to suspect that these expectations will be met in some measure. The demonstration effect for foreign firms has been significant. FDI has grown rapidly since 1982 for the rest of China, and many of the structural reforms implemented first in the SEZs have now been extended to the rest of China[33]. The price reforms now underway in Shenzhen are slowly being extended to the whole Chinese economy; the monetary reform in Shenzhen is the forerunner of currency conversion reforms in the national economy as well. The enterprise management reforms announced for the country are also well underway in many of the joint ventures and co-operative ventures within the SEZs.

PROSPECTS FOR THE SEZS

The SEZs, by both Chinese and foreign estimations, must move into a more market regulated environment[34]. This will mean rationalising the Chinese participation in joint ventures in such a way that input costs reflect real scarcities. It will also mean improving the export performance of the SEZs, once they have become fully operational.

Some senior economic analysts have foreseen the future evolution of the SEZs in the overall development scheme of the country[35]. For them, the SEZs should not be used as a model for the rest of the country. The conditions prevailing in the zones are particular to the specialised functions they were created for, and cannot be spread to the rest of China. Perhaps more surprisingly, the same sources feel that China must abandon the narrow idea "of utilising foreign capital for national construction". The country should rather, they argue, create the conditions for "all around capital co-operation with foreign countries", creating for China a place in the regional economy with balanced relationships with Japan, ASEAN and the United States. This would create a "buffer zone" for China in the region.

The SEZs are also ill-equipped to attract large transnational corporations. If transnationals can be divided broadly into three categories, namely resource developing, those seeking new markets and those seeking to transfer production facilities, the SEZs are poor choices compared to the industrial zones of Shanghai, Tianjin and northern China in general where there is a skilled labour pool, direct access to the domestic Chinese market, and relatively good infrastructure in place[36].

If the example of other Asian processing zones can be used, the scale of technology transfers as well as the level of technology transfer will be low. The backward linkages into the domestic economy from the export processing zones have been weak in other countries, and if the SEZs are isolated in the same manner from the domestic economy of China, the same may well occur in the PRC.

A possible new direction for evolution of FDI in China is in fact away from the SEZs to the more advanced technological centres in the north. Following the example of post-war Japan, China might seek to concentrate new foreign direct investment in those areas of the country where new technologies can best be assimilated. It is only there that technology innovation can really take place, where research and development capacity exists on a larger scale. The incentive for foreign firms to enter China with new technologies is access to raw materials and access to the domestic market[37].

In the context of the overall development of the country, it has been suggested that Shanghai and Guangzhou play the motor role for economic development in China; this role is and will be a regional role within the country itself. Shanghai, with its industrial base, and Guangzhou and the Pearl River Delta with its agricultural export base, must also become financial centres for China. This financial role for Shanghai at least, is also encouraged by the senior members of the Chinese banking community.

This regional "pole" theory for development was also announced as one that the leadership in Beijing hoped to see put into practice in the future[38]. The SEZs will remain small, but important centres for foreign direct investment in China; more importantly, they will act as catalysts for structural economic reforms within the country. Provided that the

SEZs are able to export enough of their goods to maintain foreign exchange reserves, they will certainly remain in the forefront of Chinese economic development. The proof of their success has been the very conditions that may lead to their marginalisation:

"In a sense, these new ways of thinking about foreign direct investment reflect the success of the zone strategy rather than its failure. It is doubtful whether a more ambitious opening to foreign capital could have been easily entertained in 1979. Since the zones were conceived, localities have demanded the right to extend SEZ-like terms to foreign investors while domestic reforms have removed some of the structural barriers impeding foreign investment. The underlying change may be that along with the domestic reforms, China's steadily growing experience with foreign investment, of which the zones have been an important part, has created a climate in which the opportunities and the advantages of a broader degree of co-operation with foreign business is becoming much more apparent"[39].

NOTES AND REFERENCES

1. See the critical article by Chen Wenhong, "What are Shenzhen's problems" (in Chinese) in *Guangjiao jing,* Hong Kong, No. 149, 1984, pp. 48-55. This article reviews the performance of Shenzhen in a critical light, and presents a number of arguments supporting the position that the SEZ has not really made as much progress as the Chinese would have Western observers believe. Those arguments are discussed in the present text. Tables for macroeconomic performance of Shenzhen are largely taken from this article.

2. Chen Wenhong, *op. cit.*

3. Chen Wenhong, *op. cit.*

4. See Chapter 3 of present study.

5. Youngson, *op. cit.,* p. 97-99.

6. K.Y. Wong, *op. cit.,* p. 19.

7. As reported by Sen-dou Chang in his unpublished paper entitled "Shenzhen SEZ: A Geographical Perspective", presented to the first Midwest Regional Seminar on China, 1982.

8. It is useful to recall the summary of the World Bank *Report on the Chinese Economy,* (Washington D.C., 1983), Vol. III, *The Social Sectors,* p. 145: "The professional knowledge in industry and agriculture is sometimes insufficient and outdated; workers often lack skill training. Labour mobility is low, the methods of hiring new staff or transferring employees from one enterprise to another are cumbersome. The immobility of the labour force may partially explain pockets of labour surplus and unexpected shortages." Journalists have presented the system of the "danwei" from the experience side, and it is more striking in its effects when told by an individual worker. See for instance, Fox Butterfield's account in *Alive in a Bitter Sea,* New York, 1983.

9. As reported in Sen-dou Chang, *op. cit.,* p. 13.

10. In *Wen Wei Po,* 7th January 1983 as reported in *SWB,* 12th January 1983.

11. Louven, *op. cit.,* p. 691; *Guide to Investment, Shenzhen.*

12. Twenty-three PRC enterprises shared eighteen industrial projects in Shenzhen in the beginning of 1982. *Xinhua,* English Press Release, 17th March 1982. This number rose to 500 by 1983.

13. See *Xinhua,* English News Release, 13th May 1983, and *SWB,* 25th May 1983. Also, Louven, *op. cit.,* p. 691.

14. Witness the description in the official publication of the *Shenzhen Special Economic Zone* (Presentation to Investors, Shenzhen, 1983): "The SEZs building (i.e. construction) workforce is made up of more than 100 000 persons coming from Beijing, Shanghai, Guangzhou, Wuhan, Chendu, and other major cities. The SEZs construction projects are carried on by close to 50 surveying, designing and building units and consultancy firms. Importance is also attached to the training of competent personnel. Special training programmes are undertaken by the Zhongshan (Sun-Yat Sen) University, Jinan University, and the Huanan University (South China Normal University in Guangzhou). For its own part, Shenzhen also has a university (Shenzhen University), a SEZ TV university, training centres of agriculture and technology in addition to special schools of construction, commerce, foreign languages, and accounting.", p. 32.

15. It is reported that between 1979-1980 more than 7 000 students dropped out of junior high school in the zone (total enrolment, 11 000). In 1981, only 83 students took part in the national competitive examination for entrance into the prestigious universities (Chongdian daxue), down from 172 in 1979. This is probably due to the creation of the SEZ and the new employment opportunities, but it is also due to the better pay and prospects for individuals in the new enterprises in the SEZ. The provincial government of Guangdong has transferred 200 teachers to the SEZ, but there are no available statistics on the number of teachers that have gone to work in the commercial sector of the SEZ. See Chang, *op. cit.* p. 14. Chang reports that only 40 per cent of the high school teachers in 1981 were college graduates, the usual minimum requirement for such a post.

16. *Shenzhen tequ bao* (Shenzhen Special Zone Herald), 10th March 1983, (in Chinese). For 1984 figures, see *SWB* FE/W1317/A/10, 12th December 1984 for comparison. The average industrial wage for the PRC in 1979 was Rmb 668 and Rmb 826 in 1984.

17. These problems were in some ways previewed in Zheng, Chen and Wei (1981) "Some Population Problems in the Construction of the Shenzhen Zone", *Jingji tequ dili Wenji* (Collected Essays on Geography of Special Economic Zone), Vol. 1, Zhongshan University, Guangzhou, pp. 20-24 (in Chinese). The above problems are also treated at length in the article "Planned Population and Labour Introduction in Shenzhen, in D. Chu, (Ed), *Shenzhen: The Largest Special Economic Zone of China*, pp. 39-47, Hong Kong, 1983, by Y.T. Ng.

19. As reported in Louven, *op. cit.*, a dynamic worker in the Nantou clothing plant can earn up to Rmb 250 a month under the piece wage system, an elasticity in income capacity that does not exist elsewhere in China. Taken from Tang Huai, "Wages in Shenzhen SEZ", *Jingji Yanjiu*, No. 6, June 1981, pp. 62 ff. (in Chinese).

20. *Beijing Information*, 26th November 1984.

21. The visits, which included those of Deng Xiaoping and other senior government officials, culminated in the visit of 120 members of the Communist Party Political Consultative Committee. This amounted to an inspection tour. The Committee went on record as supporting the SEZ opening, declaring that having seen it for themselves, "reform leads to prosperity. Reform is convincing". For an account of that visit, see *Xinhua*, Domestic Service, 19th November 1984. At the end of November 1984, Premier Zhao Ziyang visited the SEZs and openly praised the experiment, encouraging the SEZ authorities to continue in the same direction. See *Ta Kung Pao*, 29th November 1984.

22. More than 80 per cent of the new pledged investment in Shenzhen was reported to be for industrial contracts. No breakdown has yet been given for the origin of this investment.

23. See the Chen Wenhong, "Problems of a New Shenzhen Currency", *Guangjiao jing*, Hong Kong, May 1985, pp. 40-46.

24. The principal incentive is access to the domestic market for firms that have foreign currency accounts. This requirement virtually ensures that few firms will be able to see massively on the domestic market. Occasionally, a firm is set up to carry out import substitution activities in the SEZs, as the American-owned Leese Chemical Company which signed a contract with the SEZ of Shenzhen for a joint venture to produce additives to raw materials of the electroplating industry. Under the terms of the contract, the firm was given the right to market 70 per cent of the

production on the domestic Chinese market. See *SWB*, FE/W1317/A/12, 12th December, 1984.

25. "Profits and Losses of Sino-Foreign Joint Ventures, Co-operative Enterprises and 100 per cent Owned Foreign Enterprises in Shenzhen", in *Jingji daobao* (Economic Reporter), Hong Kong, 9th September 1985.

26. Another survey, with similar proportions of profit for the sample survey results, was published in *Guangdong jinrong yanjiu* (Financial Research of Guangdong), No. 2, 1984.

27. For an early review of this problem, see "The Reinstatement of Economics in China Today", C. Lin, in *China Quarterly*, 85 (March 1981).

28. See Ho and Hueneman, *China's Open Door Policy*, University of British Columbia Press, 1984, especially pp. 196 ff. for a detailed discussion of the problems with using the rate of return criterion for project evaluation.

29. As reported in the *Financial Times*, 13th May 1985.

30. "Jingji tequ de chuangkuo zuoyong" (The window role of the SEZs), in *Renmin ribao*, 8th March 1985, p. 5.

31. Gu Mu, the State Councillor in charge of the SEZs remarked in a declaration on the SEZs: "... Since the two provinces were opened to the outside world, they have conducted many experiments on economic and structural reform and accumulated some experience which plays an experimental and demonstrative role in the overall reform of the economic structure in China ... Guangdong and Fujian provinces should continue their "special policies and flexible measures" and do a still better job as experience-seeking explorers and pioneers in reform and in opening to the outside world." *SWB*, FE/7856/BII/5, 23rd January 1985.

32. *Ibid.*

33. *Ibid.* "For example, in building and managing construction projects, Shenzhen has achieved very good results in applying the method of choosing the best designs and in implementing the measures of public bidding and construction contracts. These experiences have already been popularised in the building industry across the country. In the meantime, Shenzhen has reformed its labour and personnel systems. It has put into effect the system of labour contracts floating wages, and selection and recruitment of leading cadres".

34. Even official Chinese publications have criticised Shenzhen for not becoming a totally market regulated economy: "The Special Economic Zones cannot rely on state foreign exchange support indefinitely; their operations should be directed to the needs of the international market, not the domestic Chinese market" where subsidies and price distortions exist. *Guoji shangbao*, Beijing, 25th April 1985.

35. See the remarks of Bo Tao, a senior economist with the People's Bank of China Financial Research Office in March issue of *Guangzhou Yanjiu*. This article is summarised by R. Delfs in the *Far Eastern Economic Review*, 9th May 1985, "Changing the Pattern", pp. 70-71.

36. Bo Tao, *op. cit.* "If we cannot offer substantial capacity internal markets, we will fail to attract many transnational corporations, particularly large- and medium-sized ones."

37. Cf. Chen Wenhong, *op. cit.*; Spinanger, *op. cit.*

38. Gu Mu pointed out in the above-cited speech (*SWB*, 23rd January 1985): "Our implementation policy of opening to the outside world is undergoing a constant process of summing up experience and making gradual development in the course of practice. The general trend of our opening to the outside world is a gradual development from south to north, from east to west, from coastal regions to the interior. We adopt such a step because the coastal regions have easier access to foreign markets, better communications with the outside world, a certain industrial foundation, and a more specialized personnel and management experience. Such a step conforms to China's realities and meets the needs of our economic development".

39. R. Delfs "Changing the Pattern", *Far Eastern Economic Review*, 9th May 1985, pp. 70-71.

BIBLIOGRAPHY

1. Periodicals and Newspapers

Chinese language periodicals give titles of articles in English translation, instead of romanisation.

Beijing Review, Beijing, Weekly.
China Aktuell, Hamburg, Monthly.
China Daily, Beijing, Daily.
China Economic News, Hong Kong, Weekly.
China Market, Beijing, Monthly.
China Newsletter, Tokyo, Quarterly.
China Reconstructs, Beijing, Monthly.
China Report: Foreign Broadcast Information Service, Springfield, Virginia, Daily.
China Trade Report, Hong Kong, Monthly.
Chine Commerce Extérieur, Beijing, Monthly.
Far Eastern Economic Review, Hong Kong, Weekly.
Financial Times, London, Daily.
Guangming ribao (Bright Light Daily), Beijing, Daily.
Guoji maoyi (International Trade), Beijing, Monthly.
Intertrade, Hong Kong, Monthly.
Jingji daobao (Economic Report), Hong Kong, Weekly.
Jingji ribao (Economic Daily), Beijing, Daily.
Jingji yanjiu (Economic Research), Beijing, Monthly.
Le Monde, Paris, Daily.
Ming bao (Brightness Daily), Hong Kong, Daily.
Remin ribao (People's Daily), Beijing, Daily.
Shijie jingji (World Economy), Beijing, Monthly.
Shijie jingji daobao (World Economic Herald), Shanghai, Weekly.
Summary of World Broadcasts: The Far East, London, Daily.
Summary of World Broadcasts: Weekly Economic Report, London, Weekly.
Ta Kung Pao, Hong Kong, Weekly.
The China Business Review, Washington, D.C., Bimonthly.
Wen hui bao, Hong Kong, Daily.
Xinhua yuebao (Xinhua Monthly), Beijing, Monthly.
Zhonghua renmin gongheguo guowuyuan gongbao (Bulletin of the State Council), Beijing, irregular.

2. Articles and Books

ADMINISTRATION COMMITTEE OF CHINA MERCHANTS SHEKOU INDUSTRIAL ZONE, (THE)

A Hand Book of the Administrative Bodies and the Factories and Other Enterprises in China Merchants Shekou Industrial Zone of Shenzhen SEZ of Guangdong Province, Shekou, The Administration Committee of China Merchants Shekou Industrial Zone, 1984.

AI Wei

"From Special Economic Zones to Economically-Opened Regions", *Issues and Studies*, March 1985, pp. 9-13.

ALESSANDRONI, C.

"Le zone economiche speciali nel Guangdong e nel Fujian: riflessi interni e internationali", *Mondo Cinese*, No. 33, 1981, pp. 61-77.

Almanac of China's Foreign Economic Relations and Trade 1984, Hong Kong, China Resources Trade Consultancy Co. Ltd., 1984.

ANSARI Javed, BALLANCE Robert and SINGER Hans W.

The International Economy and Industrial Development: The Impact of Trade and Investment on the Third World, Great Britain, Wheatsheaf Books Ltd., 1982.

BANK OF CHINA

Bank of China Annual Report 1983, Beijing, Bank of China, 1983.

BARNETT, A. Doak

China's Economy in Global Perspective, Washington, D.C., Brookings Institution, 1981.

BEJA Jean-Philippe

"Chine : 35 ans, ça suffit ?", *Esprit*, April 1985.

BHALLA, A.S.

Economic Transition in Hunan and Southern China, London and Basingstoke, The Macmillan Press Ltd., 1984.

BILLGREN Boel and SIGURDSON Jon

An Estimate of Research and Development Expenditures in the People's Republic of China in 1973, Paris, OECD, 1977.

BROWN, D.

"Sino-Foreign Joint Venture: Contemporary Developments and Historical Perspectives", *Journal of Northeastern Asian Studies*, December 1982.

"Bureaux des banques en Chine : réglementations provisoires", *Economie et Commerce*, February 1983, p. 7.

BUTTERFIELD Fox

Alive in Bitter Sea, New York, Times Book, 1983.

CHANG Chenpang

"Economic Conditions and Reforms in Mainland China", *Issues and Studies*, vol. XXI, No. 4, April 1985, pp. 4-8.

CHANG Ching-wen

"Opening of 14 Mainland cities to Foreign Investment", *Issues and Studies*, vol. XX, No. 5, May 1984, pp. 8-10.

CHANG, C.Y.

"Overseas in China's Policy", *China Quarterly*, No. 82, June 1980, pp. 281-303.

CHEN Erjin

China: Crossroad Socialism, (Translated by R. Robin), London, Verson Edition, 1984.

CHEN, S.B. and CHEN, M.G.

"The Establishment of Xiamen SEZ and Its Prospects", *Gangao jingji* (The Economies of Hong Kong and Macao), No. 6, 1981, pp. 49-53.

CHEN Shiwai

"Brief Discussion on Introduction of Electronic Industry's Politic", *Jingji guangli* (Economic Management), No. 73, January 1985, p. 25.

CHEN Wenhong

"Problems of a New Shenzhen Currency", *Guangjiao jing* (Wide Angle), Hong Kong, May 1985, pp. 40-46.

"What are Shenzhen's Problems?", *Guangjiao jing (Wide Angle)*, Hong Kong, April 1985, pp. 40-45.

CHEN Yongshan and WANG Muheng

"Industrial Production Export Zones in Asia and the Creation of the SEZs in China", *Zhongguo jingji wenti* (China's Economic Problems), No. 6, 1980, pp. 41-49.

CHEN Zhaobin

"A Tentative Discussion on Market Regulatory and Price Administration in the SEZs", *Zhongguo jingji wenti* (China's Economic Problems), April 1983, pp. 35-38.

CHEN Zhongjing *et alia*

La Chine et le monde, Beijing, Beijing Information, 1983.

CHEN, WEI and ZHENG

Jingji tequ dili wenti (Collected Essays on the Geography of Special Economic Zones), Guangzhou, Zhongshan University, October 1981.

CHENG Chu-yuan

China's Economic Development: Growth and Structural Change, Colorado, Westview Press, 1982.

CHEVRIER Yves

"Chine : Que veut Deng Xiaoping ?", *Politiques étrangères*, January 1985.

CHINA COUNCIL FOR THE PROMOTION OF INTERNATIONAL TRADE, DEPARTMENT OF PUBLIC RELATIONS

China's Foreign Trade Corporation and Organizations, Beijing, China Council for the Promotion of International Trade, 1982.

"China Tax Rules on Imports and Exports for Co-operation: Exploration of Offshore Oil", *Economic Review*, No. 5, 1982, pp. 22-30.

China Under the Four Modernizations (Selected Papers submitted to the Joint Economic Committee Congress of the United States), Washington, D.C., U.S. Government Printing Office, 1982.

China's Economy in the 1980's, Economic Information and Agency, 1980.

China's Foreign Economic Legislation, vol. 1, Beijing, Foreign Language Press, 1982.

China's New Economic Development Trends, Hong Kong, Economic Information and Agency, 1980.

"Chine : solution pour une crise, la réforme de la gestion dans l'entreprise chinoise", *Revue Française de Gestion*, special issue, Winter 1982/1983.

"Chine et URSS : les limites de l'ouverture économique à l'Ouest", *Économie Prospective Internationale,* No. 7, July 1981, Paris, La Documentation Française, 1981.

"Chinese Official Sees Economic Crisis", *Washington Post,* 15th June 1979.

CHOSSUDOVSKY, M.

"China's Free Trade Zones", *Co-Existence,* No. 20, 1983, pp. 41-55.

Chu goku enkaku ju-ichi sho shi no keizai (The Economic Situation of Eleven Coastal Provinces and Cities in China), Tokyo, Jetro, 1983.

Chu goku no keizai kaihatsu ku shisatsu hokoku – Shanhai, Nantsu, Nimpo, Onshu (A Study of the Economic Developing Zones – Shanghai, Nantong, Ningpo, Wenzhou), Tokyo, Nit-chu keizai kyokai, 1984.

CHU David, K.Y.

"Shekou of East Guangdong – SEZ Shantou", *Guangjiao jing* (Wide Angle), No. 143, 16th August 1984, pp. 25-29.

CHU David, K.Y. (Ed.)

Zhongguo zuida de jingji tequ – Shenzhen (The Largest Special Economic Zone of China – Shenzhen), Hong Kong, Guangjiao jing, 1983.

CHUNG Mou

"Coastal Cities and Developing Economic Zones", *Zhonggong yanjiu* (Studies on Chinese Communism), July 1984, pp. 68-75.

"Circular Concerning the Levy of Tax on Foreign Investors Who Mandate Chinese Firms as a Marketing Agency, or to Set Up Maintenance Service Supplying Spare Parts and Fitting", *Caizheng* (Finance), No. 12, 1983, p. 50.

CLARKE Christopher

China Business Manual 1982, Supplement, Washington, D.C., The National Council for US-China Trade, 1982.

"Les crédits consentis à la Chine par ses principaux partenaires en 1981-82", *Économie et Commerce,* November 1982, p. 5.

CURRIE, J.

Investment: The Growing Role of Export Processing Zones, London, EIU, 1979.

DERNBERGER Robert, F. (Ed.)

China's Development Experience in Comparative Perspective, Cambridge, Mass., London, Harvard University Press, 1980.

DERNBERGER Robert, F.

"Mainland China's Economic System: A New Model or Variations on An Old Theme?", *Issues and Studies,* Vol. XXI, No. 4, April 1985, pp. 44-72.

DEVRIES and GODEREZ

Export Processing Zones (World Bank Occasional Paper), Washington, D.C., World Bank, No. SEC M78-612.

Direct Investment in China's Four Special Economic Zones, London, Economic Intelligence Unit, No. 2, 1983, p. 19.

Doing Business with China, Washington, D.C., US Department of Commerce, 1980.

DOLAIS Yves

"45 sociétés mixtes", *Économie et Commerce,* March 1983, p. 14.

DÜRR Heiner and WIDMER Urs

Provinzstatistik der Volksrepublik China, Hamburg, Mitteilungen des Instituts für Asienkunde, 1983.

EBASHI Masahiko

Role of China in the Economic Interdependence of Asean and the Pacific, Bangkok, United Nations Economic and Social Commission for Asia and the Pacific, 1982.

ECKSTEIN Alexander (Ed.)

Quantitative Measures of China's Economic Output, Ann Arbor, The University of Michigan Press, 1980.

ECKSTEIN Alexander

China's Economic Development: The Interplay of Scarcity and Ideology, Ann Arbor, The University of Michigan, 1975.

EGAL, M. and LIU, R.

"China's SEZs", *Euro-Asia Business Review,* Vol. 2, No. 1, 1981, p. 24.

ELLIS, S.

"Decentralization of China's Foreign Trade Structures", *Georgia Journal of International and Comparative Law,* Vol. XI: 22, 1981.

ELVIN, M.

The Patterns of the Chinese Past, London, Eyre Methuen, 1973.

FABRE Guilhem

"Les zones économiques spéciales en Chine", *Le Courrier des Pays de l'Est,* No. 272, April 1983, pp. 34-42.

FAIRBANK, J.K. (Ed.)

The Chinese World Order, Cambridge, Mass., Harvard University Press, 1968.

FAIRBANK, J.K.

Trade and Diplomacy on the China Coast: The Opening of the Treaty Ports 1842-1854, Stanford, Calif., Stanford University Press, 1969.

FALKENHEIM, V.

Chinese Trade Policy, mimeo: Paper presented to Conference on Emerging Pacific Community Concept, Georgetown University, 24th-26th October 1983.

Fifth Session of the Fifth National Peoples Congress, Beijing, Foreign Language Press, 1983.

FITTING George

"Export Processing Zones in Taiwan and the PRC", *Asian Survey,* Vol. XXII, August 1982, pp. 732-44.

Free Zones in Developing Countries Expanding Opportunities for the Private Sector, Washington, D.C., USAID, November 1983.

FUJIAN INVESTMENT AND ENTERPRISE CORPORATION

Fujian Investment & Enterprise Corporation Annual Report 1983, Fujian, Fujian Investment & Enterprise Corporation, (n.d.).

GANIERE Nicole

The Process of Industrialization of China, Paris, OECD, 1973.

GARDNER John

Chinese Politics and the Succession to Mao, London and Basingstoke, The Macmillan Press Ltd., 1982.

GERAEDTS Henry, N.

The People's Republic of China: Foreign Economic Relations and Technology Acquisition 1972-1981, Sweden, University of Lund, 1983.

GERMIDIS, D. et alia

Export Processing Zones, Paris, OECD, 1984.

GERMIDIS, D. and MICHALET, C.A.

International Banks and Financial Markets in Developing Countries, Paris, OECD, 1984.

GODLEY Michael

The Mandarin-Capitalists from Nanyang: Overseas Enterprise in the Modernisation of China 1893-1911, London, Cambridge University Press, 1981.

GOOSTAD, L.

"Why the Renminbi Must Be Devalued?", *Euromoney,* October 1983.

de GRANDI Michel

"Zones spéciales économiques : pourquoi, comment ?", *Économie et Commerce,* April 1982, p. 6.

GRAY, J. and WHITE, G.

China's New Development Strategy, London, Academic Press, 1982.

GUANGDONG PROVINCIAL ADMINISTRATION OF SPECIAL ECONOMIC ZONES AND WEN WEI PO

New Regulations for Guangdong Special Economic Zones, Hong Kong, Guangdong Provincial Administration of Special Economic Zones and Wen Wei Po, 1982.

Guide to Investment, Hong Kong, Economic Information and Agency, 1982.

HO and HUENEMANN

China's Open Door Policy, The Quest for Foreign Technology and Capital, Vancouver, University of British Columbia Press, 1984.

HONG KONG AND SHANGHAI BANKING CORPORATION

Shenzhen: The Special Economic Zones of the PRC, Hong Kong, Hong Kong and Shanghai Banking Corporation, 1983.

Business Profile Series: The People's Republic of China, Hong Kong, The Hong Kong and Shanghai Banking Corporation, 1983.

HONG KONG AND SHANGHAI BANKING CORPORATION AND DEVELOPMENT COR-PORATION OF XIAMEN

Investment Guide to Xiamen SEZ, Hong Kong, Hong Kong and Shanghai Banking Corporation and the Construction and Development Corporation of Xiamen, 1984.

"Hong Kong and the SEZs", *China News Analysis,* No. 1233, 21st May 1982.

HOOKE, A.W. (Ed.)

The Fund and China in the International Monetary System, Washington, D.C., International Monetary Fund, 1983.

HOWE Christopher and KUEH, Y.Y.

"China's International Trade: Policy and Organization Change and their Place in the Economic Readjustment", *China Quarterly,* No. 100, December 1984, p. 813.

HOWE Christopher (Ed.)

Shanghai: Revolution and Development in An Asian Metropolis, Cambridge, London, New York, Rochelle, Melbourne, Sydney, Cambridge University Press, 1981.

HSIA Ronald

The Entrepôt Trade of Hong Kong with Special Reference to Taiwan and the Chinese Mainland, Taipei, Chung-hua Institution for Economic Research, 1984.

HSIAO Liang-lin

China's Foreign Trade Statistics 1864-1949, Cambridge, Mass., Harvard University Press, 1974.

HSU Immaunel C.Y.

The Rise of Modern China, New York, Oxford University Press, 1983, Third edition.

HU Yaobang

The Radiance of the Great Truth of Marxism Lights Our Way Forward, Beijing, Foreign Languages Press, 1983.

IMAI Sujikore

"Chu goku ni o keru gô ben jigyû no genjû to kadai" (Problems and Situation of the Joint Venture in China), *Chu goku keizai* (China's Economy), June 1984, pp. 43-76.

Investor's Handbook, Shenzhen, China Merchants Shekou Industrial Zone Shenzhen SEZ of Guangdong Province of PRC, 1983.

JAPAN EXTERNAL TRADE ORGANISATION

China: A Business Guide, Tokyo, Japan External Trade Organisation, 1979.

JI Chongwei

"China's Utilization of Foreign Funds and Relevant Policies", *Chinese Economic Studies,* Vol. XVII, No. 2, Winter 1983-84, pp. 37-50.

JIANG Yiwei

"Some Opinions on Shanghai Economic Development Strategy", *Jingji guanli* (Economic Management), No. 72, September 1984, pp. 12-16.

JIN Zhiguo

"A Brief of Lenin's Theories of the Development of Foreign Economic Relations and the System of Concession, and the Practice of the Special Economic Regions in China", *Jingji wenti tansuo* (Inquiry into Economic Problems), No. 36, May 1984, pp. 18-22.

JOHNSON Alexis, PACKARD, George and WILHELM, Alfred

China Policy for the Next Decade, Boston, Oelgeschalger, Gunn & Hain, 1984.

KANTON

Fukken sho homon daihyo dan kaken nokokusho (Mission Report of the Delegation to Guangdong and Fujian Province), Tokyo, Nitchu keizai kyokai, 1983.

KAWAI Hiroko

"Susumu tai-gai daihô seisaku (Open-door Policy in Works), *Chugoku keizai* (China's Economy), No. 223, July 1984, p. 1.

KELLEHER, T.

Handbook on Industrial Free Zones, UNIDO/ICD 31, 1976.

KLENNER Wolfgang and WEISEGART Kurt

The Chinese Economy Structure and Reforms in the Domestic Economy and in Foreign Trade, Hamburg, Verlag Weltarchiv GmbH, 1983.

KNIGHT Peter T.

Economic Reform in Socialist Countries: The Experiences of China, Hungary, Romania and Yugoslavia, Washington, D.C., The World Bank, 1983.

KRAUS Willy

Economic Development and Social Change in the People's Republic of China, New York, Heidelberg, Berlin, Springer-Verlag Inc., 1982.

KREYE, O.

"Export Processing Zones in Developing Countries", *UNIDO Working Papers on Structural Changes,* No. 19, August 1980 (UNIDO/ICIS 176).

KWAN Yiuwong (Ed.)

Shenzhen Special Economic Zones: China's Experience in Modernization, Hong Kong, 1982.

LARDY, N.

Agricultural Prices in China, Washington, D.C., The World Bank, 1983.

Agriculture in China's Modern Economic Development, Cambridge University Press, 1983.

LEMOINE Françoise

Réformes économiques et finances publiques en Chine, Paris, Centre d'Etudes Prospectives et d'Information Internationales, 1983.

LIANG Xuechu

"The Policy of SEZ Which Draws Experience Gained From Key Points to the Overall Area", *Jiushi niandai* (The Nineties), January 1985, pp. 60-62.

LIAO Jianxiang

"International Economic Relations and Our Open External Economic Policies", *Xueshu yanjiu* (Academic Research), No. 3, 1983, pp. 26-34.

LIN Cyril Chihren

"The Reinstatement of Economics in China Today", *China Quarterly,* No. 85, March 1981, pp. 1-48.

LIN Zili

"On the Production Responsibility System Which Links Income to Output by Production Quota Contracts", *Social Sciences in China,* No. 6, 1982.

LIPPIT Victor and SELDEN Mark (Ed.)

The Transition to Socialism in China, Armonk, New York, Croom Helm, London, M.E. Sharpe, Inc., 1982.

LIU William H.

"Foreign Trade Factor in a Centrally Planned Economy", *Issues and Studies*, Vol. XXI, No. 4, April 1985, pp. 73-107.

LO Changren

"Some Experience of Urban Development in Shenzhen Economic Special Zone", *Qiye guanli* (Business Management), 8th August 1984, pp. 32-34.

LU, L.

"The Recent Development of Shenzhen SEZ", *Gangao jingji* (The Economies of Hong Kong and Macao), No. 4, 1981, pp. 49-51.

LUO Keng and ZHANG Jiefeng

"Conversation with Liang Xiang on Shenzhen", *Baixing* (Common People), No. 91, 1st March 1985, pp. 3-6.

"Provisional Regulation Concerning the Levy of Unified Tax on Commerce and Industry, and Enterprise Revenue Tax on Foreign Contractor on Construction and Foreign Labour Service", *Caizheng* (Finance), No. 9, 1983, p. 46.

"Revitalize Enterprises and Grant the Necessary Autonomy to Them – A Seminar on the Topic of 'Loosening Ropes'", *Qiye guanli* (Business Management), June 1984, pp. 11-13.

MA Hong

New Strategies for the Chinese Economy, Beijing, New World Press, 1983.

"Major US Companies Move into Shanghai", *Economic Intelligence Unit*, No. 2, 1983, p. 19.

MORSE, H.B.

The Chronicles of the East India Company Trade to China, 1635-1843, Oxford, Clarenden Press, 1926-29.

International Relations of the Chinese Empire, London, Watson and Viney, 1930.

MORSE Ronald A. (Ed.)

The Limits of Reform in China, Boulder, Colorado, Westview Press, 1983.

MOULDER Frances

Japan, China and the Modern World Economy: Toward a Reinterpretation of East Asian Development ca 1600 to ca 1918, Cambridge, Cambridge University Press, 1977.

NEW CHINA NEWS PHOTO Co.,

China Official Annual Report 1982/83, Hong Kong, Kingsway International Publications, 1982.

OMAN Charles (Ed.)

New Forms of International Investment in Developing Countries, The National Perspective, Paris, OECD, 1984.

PAIRAULT, T.

Politique industrielle et industrialisation en Chine, Paris, La Documentation Française, October 1983.

People's Republic of China Year Book 1983, Beijing, Hong Kong, Xinhua Publishing House, 1983.

POLICY RESEARCH DEPARTMENT, MINISTRY OF FOREIGN ECONOMIC RELATIONS AND TRADE, CHINA

Guide to China's Foreign Economic Relations and Trade: Import-Export, Hong Kong, Economic Information and Agency, 1984.

Guide to China's Foreign Economic Relations and Trade: Investment Special, Hong Kong, Economic Information and Agency, 1983.

Guide to China's Foreign Economic Relations and Trade, Hong Kong, Economic Information and Agency, 1985.

"Premier bilan des ZES", *Courrier des Pays de l'Est,* No. 246, December 1980.

"Productivity, Incentive, and Reform in China's Industrial Sector", paper delivered to the 36th annual meeting of the Association of Asian Studies, Washington D.C., March 1984.

PRYBYLA Jan, S.

The Chinese Economy: Problems and Policies, University of South Carolina Press, 1981.

"Mainland China's Special Economic Zones", *Issues and Studies,* Vol. XX, No. 9, September 1984, pp. 31-50.

PYE Lucien

Chinese Commercial Negotiating Style, Santa Monica, Calif., Rand Corporation, 1982.

RAWSKI Thomas G.

Economic Growth and Employment in China, New York, Oxford, London, Oxford University Press, 1979.

Retrospect and Outlook on the Foreign Economic Relations and Trade with Fujian Province (The), Fujian, The Commission for Foreign Economic Relations and Trade, July 1983.

REYNOLDS Paul

China's International Banking and Financial System, New York, Praeger, 1982.

RICCI Matthieu and TRIGAULT Nicolas

Histoire de l'expédition chretienne au Royaume de la Chine, Reprint of 1618 edition, Paris, 1983.

SHENZHEN MUNICIPAL INDUSTRIAL DEVELOPMENT SERVICE Co.

Guide to Industrial Investment in the Near Future in Shenzhen SEZ Guangdong Province, Hong Kong, Wen Wei Po, January 1983.

SHENZHEN MUNICIPAL PEOPLE'S GOVERNMENT

Regulations and Rules of PRC on Special Economic Zones in Guangdong Province, Shenzhen, Shenzhen Municipal People's Government, 1980.

SHENZHEN SEZ DEVELOPMENT Co.

Investment Guide for the SEZs of Shenzhen, Hong Kong, Wen Wei Po, 1982.

SHI Lianrong

"Lure Foreign Capital to Set Up Industries", *Qiye guanli* (Business Management), August 1984, pp. 36-38.

SHI Yanrong and YE Xin

"Differences between Chinese and Foreign Economic Zones", *Nanfang ribao* (South Daily), 26th October 1981.

SIN Olivia

"Local Investment in Shekou Zone Over $100 million", *South China Morning Post,* 28th August 1983.

Sixth Five Year Plan of the People's Republic of China for Economic and Social Development (1981-1985), Beijing, Foreign Languages Press, 1984.

169

SPENCE Jonathan

The Gate of Heavenly Peace: The Chinese and Their Revolution, 1895-1980, New York, Viking Press, 1981.

SPINANGER, D.

"Objectives and Impact of Economic Activity Zones: Some Evidence from Asia", *Weltwirtschaftsliches Archiv*, No. 1, 1984, p. 64.

STATE STATISTICAL BUREAU, PRC

Statistical Yearbook of China 1981, Hong Kong, Economic Information and Agency, 1982.

Statistical Yearbook of China 1983, Hong Kong, Economic Information and Agency, 1983.

Statistical Yearbook of China 1984, Hong Kong, Economic Information and Agency, 1984.

STEFANI Giorgo

"Prospettive della planificazione economica pragmatica in Cina", *Mondo Cinese*, Vol. XI, No. 1, March 1983.

STEIN Daniel D. and VERZARIU Pompiliu

Joint Venture Agreements in the People's Republic of China, US Department of Commerce, 1982.

SU Wenming (Ed.)

Economic Readjustment and Reform, Beijing, Beijing Review, 1982.

SUN Ru

"The Role and the Construction of SEZs From a Strategic Point of View", *Xueshu yanjiu* (Academic Research), No. 4, 1982.

SUN Xiangjian

"The Question of the Profitability of China's Foreign Trade to the National Economy", *Social Sciences in China*, No. 3, 1983, pp. 35-60.

"Tableau complet des joint ventures", *Courrier des Pays de l'Est*, No. 272, April 1983.

TANG Huai

"Wages in Shenzhen SEZ", *Jingji yanjiu* (Economic Research), No. 5, June 1981, p. 62.

TANG Huozhao

"Experience and Analysis of Rapid Agricultural Economic Development of Shenzhen SEZ", *Jingji yu guanli yanjiu* (Research on Economic and Management), June 1984, pp. 39-43.

TENG Chao

"Brilliant Business Achievements of the Export Processing Zones", in *Jiagong chukouchu jian xun* (Export Processing Zone Concentrates), Taibei, November 1980.

"The Take-off of the Industrial Zone of Shekou", *Xueshu yanjiu* (Academic Research), No. 1, 1982.

THOMAS Stephen

Foreign Investment and China's Economic Development 1870-1900, Colorado, Westview, (n.d.).

TOBARI Higashio

"Will Shenzhen SEZ last long?", *Jing bao* (Mirror), No. 83, June 1984, pp. 38-45.

TONG Gao *et alia*

Huang qing zhi gong tu (Illustrations of the Regular Tributaries of the Imperial Qing), Palace Edition, 1761.

TSAI Min-chin

"A Survey on the Chinese Communists' Reforms of Foreign Trade System", *Zhonggong yanjiu* (Studies on Chinese Communism), No. 5, 1984, pp. 82-91.

Twelfth National Congress of the CPC (The), Beijing, Foreign Language Press, 1982.

VITTAL, N. (Ed.)

Export Processing Zones in Asia: Some Dimensions, Tokyo, Asian Productivity Association, 1977.

WANG, C.H.

"The SEZs and the Employment Problems of the PRC", *Jingji tequ dili wenti* (Collected Works of the SEZs), No. 1, 1981, pp. 31-36.

WANG He

"A Joint Venture Enterprise Insisting of Reform and Bold in Making Innovation – A Study on the Shenzhen Youyi Restaurant Co. Ltd.", *Jingji yu guanli yanjiu* (Research on Economic and Management), April 1984, pp. 35-37.

WANG Ruizhong

"Construction of Shantou SEZ Speeded", *China's Foreign Trade,* April 1983, p. 10.

WANG Shiluen

"Engineering Bidding in Shenzhen", *Qiye guanli* (Business Management), 8th August 1984, pp. 34-36.

WANG Yihe *et alia*

Zhongwai hezi jingying qiye (Chinese-Foreign Shared Capital Enterprises), Shanghai, Shehui kexueyuan, April 1984.

WANG Zhiyuan

"More on the Legislative Problems of the Special Economic Areas", *Xueshu yanjiu* (Academic Research), No. 4, 1982, pp. 38-44.

"What are the Regulations of Reduction, Exemption of Tax Concerning the Foreign Investment?", *Caizheng* (Finance), No. 6, 1983, p. 46.

WEI Lin and CHAO, Arnold

China's Economic Reforms, Philadelphia, University of Pennsylvania Press, 1982.

WONG Kwan Yin (Ed.)

Shenzhen Special Economic Zone: China's Experiment in Modernization, Hong Kong, Geographical Association, 1982.

WORLD BANK

China: Recent Economic Trends and Policy Developments, Washington, D.C., World Bank, Report No. 4072 CHA, 1983.

China: Socialist Economic Development, Washington, D.C., The World Bank, Report No. 3391 CHA, 1981, 9 vols.

WORONOFF, J.

"Export Processing Zones in Asia", *Oriental Economist,* July 1984, p. 14.

WRIGHT Stanley F.

Hart and the Chinese Customs, Belfast, W.M. Mullan & Son Ltd., 1950.

WU Chifeng, WU Qunce and ZHENG Weibiao

"Problems Brought about by the Changes in the Economic Conditions of Special Economic Areas", *Xueshu yanjiu* (Academic Research), No. 4, 1982, pp. 45-51.

WU Mei-xan

"Taiwan's Export Processing Zones and Foreign Investment", *Bank of Taiwan Quarterly*, February 1971, p. 211.

de WULF, L.

"Economic Reform in China", *Finance and Development*, March 1985.

XIAO Shang

"Would the Pre-levy of Revenue on Interest of Foreign Investment Affect the Foreign Investment?", *Caizheng* (Finance), No. 4, 1983, p. 27.

XIONG Fu

"Notes on a Journey into Shenzhen SEZ", *Hong qi* (Red Flag), No. 2, 1983, pp. 27-32.

XU Dixin *et alia*

China's Search for Economic Growth: The Chinese Economy Since 1949, Beijing, New World Press, 1982.

XUE Muqiao

China's Socialist Economy, Beijing, Foreign Languages Press, 1981.

YAHUDA Michael

Towards the End of Isolationism: China's Foreign Policy After Mao, London and Basingstoke, The Macmillan Press Ltd., 1983.

Yearbook of China's Special Economic Zone, 1983, Hong Kong, Yearbook of China's Special Economic Zone Publishing Co., 1983.

Yearbook of China's Special Economic Zone, 1984, Hong Kong, Yearbook of China's Special Economic Zone Publishing Co., 1985.

YIN Ching-yao

"The Evolution of Communist China's Foreign Policy", *Issues and Studies*, Vol. XX, No. 7, July 1984, pp. 48-69.

YOUNGSON, A.J. (Ed.)

China and Hong Kong: The Economic Nexus, Hong Kong, Oxford University Press, 1983.

YU Guangyuan

"Shenzhen SEZ", *Jingji yanjiu* (Economic Research), No. 6, June 1981, p. 62.

YU Guangyuan (Ed.)

China's Socialist Modernization, Beijing, Foreign Languages Press, 1984.

YU Yimin

Creating More Favourable Investment Environment for Foreign Investors, Paper in China's Joint Venture Law Seminar, Hong Kong, November 1983.

ZAGORIA, D. "China's Quiet Revolution", *Foreign Affairs* 62, Spring 1984.

ZENG Muye

Development and the Economic Situation in SEZ in Shenzhen and Zhuhai, Guangdong, Guangdong Institute of Economics, 1982.

"Results Obtained in Foreign Investment in the SEZs of Shenzhen and Zhuhai", *Gangao jingji* (The Economies of Hong Kong and Macao), Guangzhou, 1982.

The Take-Off of the SEZ of Shantou, Guangdong, Guangdong Institute of Economics, 1982.

ZHANG Changcai

"Some Questions Concerning the Strategic Development of the Agricultural Production in the Shenzhen Special Economic Zone", *Nongye jingji wenti* (Problems of Agricultural Economics), No. 55, July 1984, pp. 17-22.

ZHANG Hanqing

"First Approaches to the SEZs in China", *Xueshu yanjiu* (Academic Research), No. 4, 1982.

ZHANG Peijiu

"Stick to Open Policy Expand Foreign Trade", *Economic Review*, No. 5, 1982, pp. 2-5.

ZHAO Yong

"Wage System in Shekou Industrial Zone", *Qiye guanli* (Business Management), 8th August 1984, pp. 39-40.

ZHAO Ziyang

"Study New Conditions and Implement the Principle of Readjustment in an All-around Way", *Hong qi* (Red Flag), No. 1, 1980, pp. 15-20.

China's Economy and Development Principles, Beijing, Foreign Languages Press, 1982.

La situation économique et l'orientation à suivre pour notre édification, Beijing, Edition en Langues étrangères, 1982.

Zhonggong shenzhen shiwei zhengce yanjiu shi, "Public Bidding of Projects in Shenzhen", *Jingji yu guanli yanjiu* (Research on Economics and Management), April 1984, pp. 38-39.

ZHOU Guo (Ed.)

China and the World, Beijing, Beijing Review, 1982.

ZHU Naixiao

"An Inquiry into the Question of Fees for Use of Land by Stated Owned Enterprises in Shenzhen SEZ", *Jingji wenti tansuo* (Inquiry into Economic Problems), November 1984, pp. 42-45.

ANNEXES

MACROECONOMIC DATA FOR THE PEOPLE'S REPUBLIC OF CHINA

	1975	1978	1979	1980	1981	1982	1983	1984	Annual rate of change 1978-1983 (%)
Population (million)	924.2	962.6	975.4	987.7	1 000.7	1 015.4	1 025.0	1 037.0	1.3
Production (1980 constant Rmb billion)									
Gross Output Value of Agriculture (GOVA)	173.5	197.0	213.9	222.3	236.9	263.2	288.2	330.0	7.9
Gross Output Value of Industry	320.6	421.4	457.3	497.2	517.8	557.7	616.4	700.1	7.9
of which: % Light Industry	43.3	42.7	43.1	46.9	51.4	50.5	49.6	49.5	–
% Heavy Industry	56.7	57.3	56.9	53.1	48.6	49.5	50.4	50.5	–
Total Gross Production (GOVIA)	494.1	618.4	671.2	719.5	754.7	820.9	904.6	1 030.1	7.9
Net Material Product (NMP)	275.0	324.0	346.6	368.8	386.9	419.0	457.1	549.0*	7.1
Expenditure on National Income (Current Rmb billion)									
Total	245.1	297.5	335.6	368.8	388.7	425.6	473.1	n.a.	–
of which: % accumulation	33.9	36.5	34.6	31.6	28.5	29.0	30.0	n.a.	–
% public consumption	6.9	7.3	8.5	8.1	7.9	7.8	7.9	n.a.	–
% private consumption	59.2	56.2	56.9	60.3	63.6	63.2	62.1	n.a.	9.7
State Budget (current Rmb billion)									
State Revenues	81.6	112.1	110.3	108.5	108.9	112.4	124.9	n.a.	+2.2
State Expenditures	82.1	111.1	127.4	121.3	111.5	115.3	129.2	n.a.	+3.1
Deficit	–0.5	+1.0	–17.1	–12.8	–2.6	–2.9	–4.3	n.a.	
External Sector (current Rmb billion)									
Exports (fob)	14.3	16.6	21.2	27.1	36.8	41.4	43.8	45.8	+21.1
Imports (cif)	14.5	18.7	24.3	29.9	36.8	35.8	42.2	45.0	+19.2
Commercial balance	–0.2	–1.9	–3.1	–2.8	0.0	+5.6	+1.6	+0.8	–
Foreign Exchange Revenues ($ billion)	n.a.	1.6	2.2	2.5	5.0	11.3	14.9	15.1	–
Money and Prices									
RPI (% change/previous year)	0.2	0.6	2.0	6.0	2.4	1.9	1.5	2.8	–
Currency in circulation (Rmb billion)	n.a.	21.2	26.8	34.6	39.6	43.9	53.0	n.a.	–
Exchange Rate/Rmb/$	1.94	1.68	1.56	1.50	1.70	1.89	1.97	2.32	

Source: Compiled by Development Centre from various national and international sources.

* At current Rmb billion.

EQUITY JOINT VENTURE IN PRC EFFECTIVE 1979-NOVEMBER 1984 BY COUNTRY/TERRITORY

Country/territory	Total No. of ventures	No. of inside the SEZ	Ventures outside the SEZ	Whole of China		Inside SEZ investment		Outside SEZ investment		Average size		Foreign share	
				$ million		$ million		$ million		$ thousand		%	
				Total investment	foreign investment	Total investment	foreign investment	Total investment	foreign investment	SEZ	Outside SEZ	SEZ	Outside
(1)	(2)	(3)	(4)	(5)	(6)	(7)	(8)	(9)	(10)	(11)	(12)	(13)	(14)
I. Joint-venture involving one foreign partner													
United States	38 (22)	8 (5)	30 (17)	834.20	446.05	58.95	29.17	775.88	416.88	5 835	24 522	50.00	53.73
Japan	56 (34)	9 (5)	47 (29)	316.52	231.04	70.61	42.28	245.91	189.08	7 758	6 668	64.45	74.53
Hong Kong	71 (57)	33 (32)	38 (25)	132.94	64.94	67.69	36.06	65.25	28.88	1 127	1 155	53.28	44.26
Fed. Rep. of Germany	3 (2)	1 (1)	2 (1)	28.19	14.09	27.49	13.75	0.69	0.34	13 748	345	50.00	50.00
Belgium	1 (1)	–	1 (1)	22.50	9.00	–	–	22.50	9.00	–	9 000	–	40.00
Singapore	3 (1)	3 (1)	–	20.00	6.00	20.00	6.00	–	–	6 000	–	30.00	–
Sweden	1 (1)	–	1 (1)	12.00	6.00	–	–	12.00	6.00	–	6 000	–	50.00
United Kingdom	4 (2)	2 (2)	2 (0)	7.00	3.77	7.70	3.77	–	–	1 886	–	49.00	–
Philippines	3 (3)	1 (1)	2 (2)	5.64	2.77	4.51	2.26	1.12	0.51	2 258	256	50.00	45.60
Denmark	1 (1)	1 (1)	–	5.01	2.50	5.01	2.50	–	–	2 506	–	50.00	–
Norway	1 (1)	–	1 (1)	2.50	1.25	–	–	2.50	1.25	–	1 250	–	50.00
Thailand	1 (1)	–	1 (1)	0.76	0.26	–	–	0.76	0.27	–	267	–	35.00
France	3 (1)	2 (0)	1 (1)	7.07	0.20	–	–	0.53	0.20	–	202	–	38.00
Australia	1 (1)	–	1 (1)	0.46	0.23	–	–	0.46	0.23	–	230	–	50.00
Switzerland	1 (1)	1 (1)	–	0.36	0.18	.36	0.18	–	–	179	–	50.00	–
Finland	1 (0)	1 (1)	–	16.46	0.07	–	–	16.46	0.07	–	659	–	4.0
Italy	1 (0)	–	1 (0)	–	–	–	–	–	–	–	–	–	–
Unspecified	6 (3)	4 (2)	2 (1)	4.42	2.84	1.52	1.39	2.9	1.45	693	1 450	91.13	50.00
II. Joint-venture involving more than one foreign partner													
Hong Kong-Others	9 (5)	3 (2)	6 (3)	189.08	60.04	50.90	25.05	138.18	34.99	12 525	11 664	49.20	25.30
US-Singapore	1 (0)	–	1 (0)	–	–	–	–	–	–	–	–	–	–
Total	206 (138)	69 (54)	137 (84)	1 601.29	849.78	314.74	162.41	1 275.16	683.36	3 008	8 141	51.60	53.63

Source: Compiled by OECD from "Delegates Survey Report on the Investment Environment in China" October 1984, Sino-Japan Trade Association.

Notes:
- Components do not add to total because of rounding.
- Amount of investment at current US dollars.
- Only ventures when both the amount of: a) total investment and b) foreign investment (or the latter's share) are available were they taken into account.
- In case only a) are available it would not be possible to arrive at figures for (9)-(12) corresponding to each venture. But one can still work out a sum of contractual total investment which would now stand at $2 171.33 million rather than $1 584.75 million as shown at the bottom of column (5).
- Of joint ventures chosen for the present coverage, those entering into present calculation appear in brackets for columns (2), (3) and (4), respectively, which serve as the sample base.
- Conversion of other currency value of investment into US dollars applies at the time when either the contract is a) concluded or b) approved or basically approved by the Chinese Authorities or c) being concluded but awaiting finalisation. When monthly average exchange rate is not available the one prevailing at the end of the month referred to or an approximate rate is used instead.
- In the case where both a maximum and minimum equity investment is agreed upon (and respectively the share of foreign holding spreads over a range) then their average is used as the basis of calculation.

Annex 3

COMPARATIVE TABLE OF INCENTIVES FOR FOREIGN INVESTMENT IN SPECIAL ECONOMIC ZONES, AND IN THE REST OF CHINA[1]

(1984. These incentives are constantly evolving)

I. TAXES

1. Income tax		
a) Compensation trade;	G	Not subject to income tax in China.
b) Processing contract;	G	Not subject to income tax in China.
c) Contractual joint ventures;	G	Foreign party will pay tax on its share of nets profits on a progressive scale between 20 per cent and 40 per cent plus a local surtax of 10 per cent of national taxes (i.e. between 2 per cent and 4 per cent of net profits).
d) Direct sale[2];	SEZ	In the Special Economic Zones there is no local surtax. Passive China-source incomes (interest, royalties, licence fees...) are subject to a 20 per cent withholding tax. However, this rate can be lowered to 10 per cent or 0 per cent for contracts involving advanced technology, technical data, technical training.
e) Equity joint venture;	G	The rate of taxation on net income[3] is set at 30 per cent plus a 3 per cent local surtax. However, companies may apply for an exemption for the two first profit-making years[4] and a 50 per cent reduction for the three following years.
	SEZ	Income tax rate is set at 15 per cent without any local surtax. Ventures commencing operations before 1985, or investing more than HK$ 5 million, using advanced technology, those having a long lead time, or considered as highly desirable, may apply for an exemption for the first three profit-making years or a reduction of 20-50 per cent (Shenzhen) this exemption can extend to up to five years (Xiamen). Investors whoseprofits are reinvested for no less than five years can apply for a tax reduction/exemption on reinvested profits.
f) 100 per cent owned companies.	G	They require a special authorisation to start businesses outside Economic or Special Economic Zones.
	SEZ	They are allowed to settle in SEZs and ETDZs. They enjoy basically the same treatment as do Joint Ventures.
2. Accelerated Depreciation	G	Straight line method is usually used. Depreciation period is between five years for electronic equipment and thirty years for buildings.
	SEZ	Faster depreciation rates can be granted to joint ventures inside the SEZ.
3. Remittance tax	G	After all taxes and legal contributions to different funds are paid. The same treatment applies for capital remitted abroad after a 10 per cent withholding tax is paid. The same treatment applies for capital remitted abroad after the termination/liquidation of a company, once all liabilities and taxes have been repaid.
	SEZ	No withholding tax.
4. Import tax	G	Rules are not yet very clear, but they should net be very different from those implemented in the SEZ if imports are used to produce export goods.
	SEZ	Investment goods and raw materials are exempt. Consumer goods can enjoy a reduction/exemption of

domestic market are subject to export tax, and to the repayment of import taxes that were not levied on the inputs incorporated.

6. Commercial and Industrial Consolidated Tax (CICT)	G	Capital contributed in the form of imported machinery equipment and spare parts, or additional capital of the same sort can be exempt from CICT, provided imported items cannot be produced in China. Enterprises experiencing difficulties in paying CICT for sales on the domestic market can apply for reduction/exemption. Enterprises may apply for exemption on CICT for export goods, except for a few commodities.
	SEZ	*i)* Construction or production imports; *ii)* A reasonable amount of office supplies; *iii)* Means of transportation imported by foreign representation for their own use; *iv)* Food and beverages imported for tourists and restaurants can be exempt from CICT upon approval. – A 50 per cent reduction of CICT available on imported high tax commodities; – If goods are manufactured mainly for export purposes, no CITC is levied at the factory level, except for a few types of commodities; – In Shenzhen, municipal authorities consider lowering the rate of CICT, when applicable.
7. Local taxes	G	Local taxes include net income surtax, real estate tax, vehicle licence tax, vessels and licence tax. They are assessed at the discretion of local authorities.
	SEZ	No local surtax is levied on net income.
8. Personal tax	G	Foreign personnel staying less than five years for their work, and with no intention of becoming permanent residents, irrespective of whether or not they remit their overseas income to China, are not required to report or pay tax on their overseas profit. They may remit freely 50 per cent of their Chinese income after tax overseas after paying a 10 per cent withholding tax and apply to the Bank of China if they wish to remit a higher proportion of their Chinese after tax income.
	SEZ	Income earned inside China will enjoy a 50 per cent tax cut. After tax income can be freely remitted without any withholding tax. Furthermore, foreign employees enjoy reduction/exemption of import tax on daily life necessities.
9. Special conditions for Overseas Chinese Investors	SEZ	In Xiamen, Taiwan investors who wish to do business in the SEZ will enjoy special preferential treatment in enterprise income tax. Furthermore, there will be no income tax levied on foreign workers in any overseas Chinese investment.

II. COSTS[5]

1. Industrial Land Rents	SEZ	Industrial land rent in Shenzhen is in the range HK$/19-32 per sq. m./year, in Shantou HK$/16-64 per sq. m./year. In the three Guangdong Zones, this rent can be revalued every three years by a rate not exceeding 30/per cent. In Xiamen rent is in the range HK$ 3-63 per sq. m./year and can be revalued every five years by less than 20 per cent.
2. Industrial standardised buildings	SEZ	The Special Economic Zones' Development Companies provide standardised office space for investors. In Shenzhen, the monthly rate range is HK$/1 200 sq. m.

Annex 3 (cont'd)

COMPARATIVE TABLE OF INCENTIVES FOR FOREIGN INVESTMENT IN SPECIAL ECONOMIC ZONES, AND IN THE REST OF CHINA[1]

(1984. These incentives are constantly evolving)

II. COSTS[5] *(cont'd)*

3. Purchase of workshops	SEZ	A company which so wishes can purchase workshops. In Shenzhen the rate is HK$/1 100-1 900 per sq. m.; it Shantou it is HK$/1 200 sq. m.
4. Costs of water and energy	SEZ	These costs are said to be lower than in Hong Kong; water: Rmb/0.18 cubic metres (HK$/2.7 per 1 000 gallons), electricity: Rmb 0.085 kwh for industrial use, Rmb 0.2/kwh for domestic use.
5. Participation to infrastructure costs	SEZ	If the company is located in places where there is no infrastructure, it will be required to pay a specified amount for construction fees.
6. Labour costs[6]	SEZ	Labour costs have three components (1984): 70 per cent is given directly to the worker; 25 per cent is used for social labour insurance, and to compensate for various state subsidies for workers, 5 per cent is reserved by the enterprise to subsidise its welfare fund. It is reported that some companies have given workers a 10 per cent bonus in foreign exchange certificates. It has further been reported that in Shekou, CMSN and foreign investors have an agreement that ensures profit to the investor; the latter are allowed to hold back part of wages to offset losses either in production or in trade. Labour costs vary from about HK$ 500 in Shantou, Zhuhai and Xiamen to around 700 in Shenzhen and 800 in Shekou. Wages are to be increased each year between 5 per cent and 15 per cent. An amount equivalent to 2 per cent of total wage bill must be given to labour unions. Enterprises may choose the system of remuneration (piecework, hourly basis, daily basis, proposition of fixed and floating wage...). They may also choose the work schedule (number of shifts). The normal standard is eight hours a day, six days a week.
7. Labour supply	SEZ	Enterprise may hire employees proposed by a local labour service company. They may also conduct their own selection tests. Labour contracts include a provision for a three to six month trial basis. Enterprises may dismiss workers. Workers are recruited locally in Zhuhai, Shantou and Xiamen. Shenzhen also recruits them nationally via joint ventures with state enterprises in the PRC. The People's Liberation Army has been used to provide workers in infrastructural projects. These workers are grantd "temporary" residence status in the SEZ.
8. Currency used for settlement	SEZ	Usually, foreign currencies are sued to pay costs described above. But if Renminbi are earned through access to the domestic market, companies may apply to use them to settle these costs.
9. Rate of change[7]	SEZ	Net hard currency earnings from exports are entitled to be changed into Renminbi at international settlement rate (2.8 Rmb/$) rather than official rate (2.5 Rmb/$ in September 1984) by the Bank of China.
10. Special conditions for Overseas Chinese Investors	SEZ	In Xiamen, Taiwan investors will enjoy special conditions for land rents.

1.	Access to domestic inputs	G	Materials needed by a joint venture should be priced according to the current prices in China and paid for in Renminbi except for precious metals, petroleum, coal, and timber that are valued at their international price and paid for in Renminbi. However, if the material falls in the category of goods imported (respectively exported) by China, it will be priced according to the CIF international prices plus import duty plus business tax plus import commission fees (respectively according to the international fob price).
		SEZ	Companies within SEZs do not usually pay tax on imported goods from abroad; therefore goods exported by China will be priced preferentially on the basis of their international fob price.
2.	Access to domestic market	G	The basic principle is: the foreign exchange balance of the company should be maintained through export sales. On a case by case basis, wider access to the domestic market can be granted to goods that are otherwise imported by China. Products should be sold to the relevant Foreign Trade Corporation at a value related to their international price, and usually paid for in foreign currency.
		SEZ	Unlike projects in the 14 coastal cities or others in the rest of China that are often renovation projects, investments in the SEZ should be export-oriented. The ratio of domestic sales (average 20 per cent) depends upon demand. However, at least one joint venture was recently allowed an 80 per cent ratio.
3.	Access to domestic finance	G	A joint venture may apply to Chinese banks for loans in Renminbi or in foreign currency for capital construction, or operational turnover. Interest rates on loans in Renminbi (respectively Foreign Currency) will be calculated in accordance with rates set by the People's Bank of China (respectively Bank of China). Typical rates in Renminbi loans are around 5 per cent, foreign currency loans range between market rates and market rate – 5 per cent owing to the current large surplus in foreign currency of China. The Bank of China can also accept to be the guarantor in the case where the foreign participant in a joint venture wishes to get loans from a foreign bank. However, the Chinese participant will not provide such a guarantee, and Chinese Government authorities are reluctant to provide such a guarantee.
4.	Special conditions for Chinese overseas	SEZ	In Xiamen, Taiwan investors will be permitted to sell at least 30 per cent of their production on the domestic market. Furthermore, they will enjoy preferential rates on the loans in Renminbi or in foreign currency borrowed from Chinese banks.

The 14 coastal cities and their respective Economic and Technical Development Zones are not included in this table.
1. Although direct sales contracts are in essence different from an investment, they are included here, since foreign companies may have to choose between these two forms of involvement.
2. Net income has been given a very precise accounting definition for joint ventures in two documents issued by the Ministry of Finance: "Accounting System for Joint Ventures with Chinese and Foreign Investment", provisional draft, Beijing, 1983 and "Chart of Accounts and Forms of Accounting Statements for Joint Ventures with Chinese and Foreign Investment", provisional draft, Beijing, 1983. Basically, internationally accepted accounting principles are adopted except that the "cost or market whichever is lower" method for inventory valuation is not used and "reserves for bad debts" are not set up.
3. The first profit making year is defined as the first year in which a joint venture realises profits after accumulated operating losses from prior years have been made up.
4. Most of the costs described in this section are set by municipal authorities, and subject to negociation. The figures given should be taken only as samples.
5. This is the official presentation of labour costs. In fact other practices are also reported, such as factories paying extra bonuses in foreign currency to make up for inflation in the SEZ.
6. This distinction between official exchange rate and internal settlement rate became obsolete in 1985 for most operations in China.
7. "G" means: applies to the whole of China unless otherwise stated. "SEZ" means: only concerns the four Special Economic Zones.
Notation: Tables compiled by the OECD Development Centre from various published and unpublished sources as well as from interviews with Chinese officials.

SELECTED DATA ON SEZs

GUANGDONG SEZ DEVELOPMENT PROGRAMME TO YEAR 2000

Category	Shenzhen	Zhuhai	Shantou
Urban population:			
1981	30 000	25 000	negligible
1985	250 000	n.a.	n.a.
1990	400 000	n.a.	28 000
1995	n.a.	87 900	n.a.
2000	1 000 000	175 000	52 000
Area (sq. km.) (1984)	327.5	6.8	1.6
Built-up	98	4.37	–
Industrial	15	2.27	1.6
Employment			
Basic	87 000	25 000	44 000
Non-basic	58 000	16 700	5 200[1]
Investment (2000) cumulative (Rmb million)			
Urban	1 670	272.5	88.4
Infrastructure	2 000	400	45
Total Chinese construction	3 670	672.5	133.4
Overseas investment ($ million)	1 500	227	280
Total industrial investment (2000) (Rmb million)[2]	3.75	n.a.	n.a.
Target number of factories (2000)	1 500	227	160
Expected output value (2000) (Rmb billion)	10^2	n.a.	n.a.
Expected export value (2000) ($ million per annum)	3 750	568	400

1. *Wen Wei Po*, 1st February 1983.
2. China Symposium, "Business Opportunities in the Guangdong Economic Zones", Hong Kong, November 1982.

Source: D. Chu, "The Costs of the Four SEZs to China", *Economic Reporter*, June 1982, p. 9.

SELECTED DATA ON SEZs *(cont'd)*

XIAMEN SEZ DEVELOPMENT PROGRAMME TO YEAR 2000

Population		
1984		329 800
1990		500 000
2000		800 000
Area (sq. km.)		129.91
Industrial		15.00
Employment (2000)		440 000
Investment (2000)		
(Rmb million)		
Urban		4 400
Infrastructure		6 600
Total		11 000
Overseas investment ($ million)		3 000
Export value ($ million)	1983	125
	2000	2 000
GVIO (Rmb million)	1983	1 350
GVIO (Rmb billion)	2000	18
GVAO (Rmb million)	1983	40
	2000	100

Source: Xiamen United Development Company.

SHENZHEN: BASIC INFRASTRUCTURAL INVESTMENT AND INDUSTRIAL OUTPUT

Unit: Rmb million

	1979	1980	1981	1982	1983	1984
Capital construction	50	125	270	633	886	1 584
GVIO	61	84	243	362	720	1 800

Source: Renmin Ribao, 29th March 1984.
State Statistical Bureau, 1985.

Annex 4
SELECTED DATA ON SEZs *(cont'd)*

CONTRIBUTION OF SHENZHEN SEZ TO THE GROWTH OF GUANGDONG PROVINCE VALUE OF INDUSTRIAL OUTPUT (GVIO)

	1980	1981	1982	1983	Total
Shenzhen SEZ					
GVIO	0	238	350	600	
Increment (A)		238	112	250	600
Guangdong					
GVIO	22 199	25 036	27 218	29 940	
Increment (B)		2 837	2 182	2 722	7 741
(A)/(B) (%)		8.4	5.1	9.1	7.8

Sources: *State Statistical Bureau,* 1983, Development Centre calculations from *Provinzstatistik der Volkrepublik,* Institut Fur Asienkunde, Hamburg, Federal Republic of Germany, 1983, and diverse published sources.

FOREIGN EXCHANGE BALANCE OF SEZs

Foreign exchange inflows	Foreign exchange outflows or earning	Foreign exchange retained

A. Gross exports
B. Foreign capital
C. Loans from abroad
	A. Payments by SEZ firms
	  1. Imported prod. goods
	  2. Imported prod. services
	  3. Repatriated salaries, dividends, royalties
	  4. Repatriated profits of firms facilities
	  5. Financing costs (interest and principal)
	
	B. Payments by host country
	  1. Import content of locally provided inputs
	  2. Imports for SEZ developments
	  3. Financing costs (1 & 2)
	  4. Overseas SEZ promotion costs
	
	C. Repatriated terminal value of foreign SEZ firms
		A. Payments made in hard currency or for which the SEZ firm has to change hard currency
		  1. Locally provided goods
		  2. Locally provided services
		  3. Wages (repat. salaries)
		  4. Land lease factory rent maintenance
		  5. Tax payments
		  6. Remittance tax
		
		B. Foreign capital investment minus repatriated terminal value

Source: Adopted from *Export Processing Free Zones in Developing Countries, Implications for Trade and Industrialisation Policies,* TD/B/C.2/211, 18th January 1983, p. 28, UNCTAD, Geneva.

MAP OF CHINA'S FOUR SEZ AND 14 COASTAL CITIES

CHINA'S SPECIAL ECONOMIC ZONES

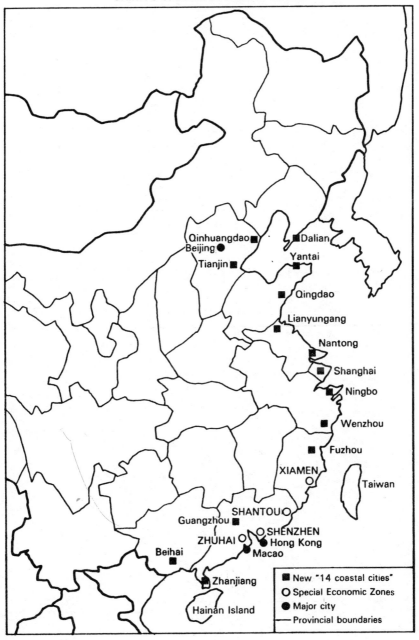

Qinhuangdao ■ ■ Dalian
Beijing ●
Yantai
Tianjin ■ ■

Qingdao ■

Lianyungang ■

Nantong ■

Shanghai ■

Ningbo ■

Wenzhou ■

Fuzhou ■

XIAMEN ○

Taiwan

SHANTOU ○

Guangzhou ■ ○ SHENZHEN
ZHUHAI ○ ● Hong Kong
● Macao

Beihai ■

Zhanjiang ■

Hainan Island

■	New "14 coastal cities"
○	Special Economic Zones
●	Major city
—	Provincial boundaries

OECD SALES AGENTS
DÉPOSITAIRES DES PUBLICATIONS DE L'OCDE

ARGENTINA - ARGENTINE
Carlos Hirsch S.R.L.,
Florida 165, 4° Piso,
(Galeria Guemes) 1333 Buenos Aires
Tel. 33.1787.2391 y 30.7122

AUSTRALIA-AUSTRALIE
D.A. Book (Aust.) Pty. Ltd.
11-13 Station Street (P.O. Box 163)
Mitcham, Vic. 3132   Tel. (03) 873 4411

AUSTRIA - AUTRICHE
OECD Publications and Information Centre,
4 Simrockstrasse,
5300 Bonn (Germany)   Tel. (0228) 21.60.45
Local Agent:
Gerold & Co., Graben 31, Wien 1 Tel. 52.22.35

BELGIUM - BELGIQUE
Jean de Lannoy, Service Publications OCDE,
avenue du Roi 202
B-1060 Bruxelles   Tel. 02/538.51.69

CANADA
Renouf Publishing Company Limited/
Éditions Renouf Limitée Head Office/
Siège social – Store/Magasin :
61, rue Sparks Street,
Ottawa, Ontario KIP 5A6
Tel. (613)238-8985. 1-800-267-4164
Store/Magasin : 211, rue Yonge Street,
Toronto, Ontario M5B 1M4.
Tel. (416)363-3171
Regional Sales Office/
Bureau des Ventes régional :
7575 Trans-Canada Hwy., Suite 305,
Saint-Laurent, Quebec H4T 1V6
Tel. (514)335-9274

DENMARK - DANEMARK
Munksgaard Export and Subscription Service
35, Nørre Søgade, DK-1370 København K
Tel. +45.1.12.85.70

FINLAND - FINLANDE
Akateeminen Kirjakauppa,
Keskuskatu 1, 00100 Helsinki 10 Tel. 0.12141

FRANCE
OCDE/OECD
Mail Orders/Commandes par correspondance :
2, rue André-Pascal,
75775 Paris Cedex 16
Tel. (1) 45.24.82.00
Bookshop/Librairie : 33, rue Octave-Feuillet
75016 Paris
Tel. (1) 45.24.81.67 or/ou (1) 45.24.81.81
Principal correspondant :
Librairie de l'Université,
13602 Aix-en-Provence   Tel. 42.26.18.08

GERMANY - ALLEMAGNE
OECD Publications and Information Centre,
4 Simrockstrasse,
5300 Bonn   Tel. (0228) 21.60.45

GREECE - GRÈCE
Librairie Kauffmann,
28 rue du Stade, Athens 132   Tel. 322.21.60

HONG KONG
Government Information Services,
Publications (Sales) Office,
Beaconsfield House, 4/F.,
Queen's Road Central

ICELAND - ISLANDE
Snæbjörn Jónsson & Co., h.f.,
Hafnarstræti 4 & 9,
P.O.B. 1131 – Reykjavik
Tel. 13133/14281/11936

INDIA - INDE
Oxford Book and Stationery Co.,
Scindia House, New Delhi 1   Tel. 45896
17 Park St., Calcutta 700016   Tel. 240832

INDONESIA - INDONESIE
Pdin Lipi, P.O. Box 3065/JKT.Jakarta
Tel. 583467

IRELAND - IRLANDE
TDC Publishers – Library Suppliers
12 North Frederick Street, Dublin 1
Tel. 744835-749677

ITALY - ITALIE
Libreria Commissionaria Sansoni,
Via Lamarmora 45, 50121 Firenze
Tel. 579751/584468
Via Bartolini 29, 20155 Milano   Tel. 365083
Sub-depositari :
Ugo Tassi, Via A. Farnese 28,
00192 Roma   Tel. 310590
Editrice e Libreria Herder,
Piazza Montecitorio 120, 00186 Roma
Tel. 6794628
Agenzia Libraria Pegaso,
Via de Romita 5, 70121 Bari
Tel. 540.105/540.195
Agenzia Libraria Pegaso, Via S.Anna dei
Lombardi 16, 80134 Napoli.   Tel. 314180
Libreria Hœpli,
Via Hœpli 5, 20121 Milano   Tel. 865446
Libreria Scientifica
Dott. Lucio de Biasio "Aeiou"
Via Meravigli 16, 20123 Milano Tel. 807679
Libreria Zanichelli, Piazza Galvani 1/A,
40124 Bologna   Tel. 237389
Libreria Lattes,
Via Garibaldi 3, 10122 Torino   Tel. 519274
La diffusione delle edizioni OCSE è inoltre
assicurata dalle migliori librerie nelle città più
importanti.

JAPAN - JAPON
OECD Publications and Information Centre,
Landic Akasaka Bldg., 2-3-4 Akasaka,
Minato-ku, Tokyo 107   Tel. 586.2016

KOREA - CORÉE
Pan Korea Book Corporation
P.O.Box No. 101 Kwangwhamun, Seoul
Tel. 72.7369

LEBANON - LIBAN
Documenta Scientifica/Redico,
Edison Building, Bliss St.,
P.O.B. 5641, Beirut   Tel. 354429-344425

MALAYSIA - MALAISIE
University of Malaya Co-operative Bookshop
Ltd.,
P.O.Box 1127, Jalan Pantai Baru,
Kuala Lumpur   Tel. 577701/577072

NETHERLANDS - PAYS-BAS
Staatsuitgeverij Verzendboekhandel
Chr. Plantijnstraat, 1 Postbus 20014
2500 EA S-Gravenhage   Tel. 070-789911
Voor bestellingen: Tel. 070-789208

NEW ZEALAND - NOUVELLE-ZÉLANDE
Government Printing Office Bookshops:
Auckland: Retail Bookshop, 25 Rutland Street,
Mail Orders, 85 Beach Road
Private Bag C.P.O.
Hamilton: Retail: Ward Street,
Mail Orders, P.O. Box 857
Wellington: Retail, Mulgrave Street, (Head
Office)
Cubacade World Trade Centre,
Mail Orders, Private Bag
Christchurch: Retail, 159 Hereford Street,
Mail Orders, Private Bag
Dunedin: Retail, Princes Street,
Mail Orders, P.O. Box 1104

NORWAY - NORVÈGE
Tanum-Karl Johan a.s
P.O. Box 1177 Sentrum, 0107 Oslo 1
Tel. (02) 801260

PAKISTAN
Mirza Book Agency
65 Shahrah Quaid-E-Azam, Lahore 3 Tel. 66839

PORTUGAL
Livraria Portugal,
Rua do Carmo 70-74, 1117 Lisboa Codex.
Tel. 360582/3

SINGAPORE - SINGAPOUR
Information Publications Pte Ltd
Pei-Fu Industrial Building,
24 New Industrial Road No. 02-06
Singapore 1953   Tel. 2831786, 2831798

SPAIN - ESPAGNE
Mundi-Prensa Libros, S.A.,
Castelló 37, Apartado 1223, Madrid-28001
Tel. 431.33.99
Libreria Bosch, Ronda Universidad 11,
Barcelona 7   Tel. 317.53.08/317.53.58

SWEDEN - SUÈDE
AB CE Fritzes Kungl. Hovbokhandel,
Box 16356, S 103 27 STH,
Regeringsgatan 12,
DS Stockholm   Tel. (08) 23.89.00
Subscription Agency/Abonnements:
Wennergren-Williams AB,
Box 30004, S104 25 Stockholm. Tel. 08/54.12.00

SWITZERLAND - SUISSE
OECD Publications and Information Centre,
4 Simrockstrasse,
5300 Bonn (Germany)   Tel. (0228) 21.60.45
Local Agent:
Librairie Payot,
6 rue Grenus, 1211 Genève 11
Tel. (022) 31.89.50

TAIWAN - FORMOSE
Good Faith Worldwide Int'l Co., Ltd.
9th floor, No. 118, Sec.2
Chung Hsiao E. Road
Taipei   Tel. 391.7396/391.7397

THAILAND - THAILANDE
Suksit Siam Co., Ltd.,
1715 Rama IV Rd.,
Samyam Bangkok 5   Tel. 2511630

TURKEY - TURQUIE
Kültur Yayinlari Is-Türk Ltd. Sti.
Atatürk Bulvari No: 191/Kat. 21
Kavaklidere/Ankara   Tel. 17.02.66
Dolmabahce Cad. No: 29
Besiktas/Istanbul   Tel. 60.71.88

UNITED KINGDOM - ROYAUME UNI
H.M. Stationery Office,
Postal orders only:
P.O.B. 276, London SW8 5DT
Telephone orders: (01) 622.3316, or
Personal callers:
49 High Holborn, London WC1V 6HB
Branches at: Belfast, Birmingham,
Bristol, Edinburgh, Manchester

UNITED STATES - ÉTATS-UNIS
OECD Publications and Information Centre,
Suite 1207, 1750 Pennsylvania Ave., N.W.,
Washington, D.C. 20006 - 4582
Tel. (202) 724.1857

VENEZUELA
Libreria del Este,
Avda F. Miranda 52, Aptdo. 60337,
Edificio Galipan, Caracas 106
Tel. 32.23.01/33.26.04/31.58.38

YUGOSLAVIA - YOUGOSLAVIE
Jugoslovenska Knjiga, Knez Mihajlova 2,
P.O.B. 36, Beograd   Tel. 621.992

Orders and inquiries from countries where Sales
Agents have not yet been appointed should be sent
to:
OECD, Publications Service, Sales and
Distribution Division, 2, rue André-Pascal, 75775
PARIS CEDEX 16.

Les commandes provenant de pays où l'OCDE n'a
pas encore désigné de dépositaire peuvent être
adressées à :
OCDE, Service des Publications. Division des
Ventes et Distribution. 2. rue André-Pascal. 75775
PARIS CEDEX 16.

69482-03-1986

OECD PUBLICATIONS, 2, rue André-Pascal, 75775 PARIS CEDEX 16 - No. 43509 1986
PRINTED IN FRANCE
(41 86 03 1) ISBN 92-64-12801-8